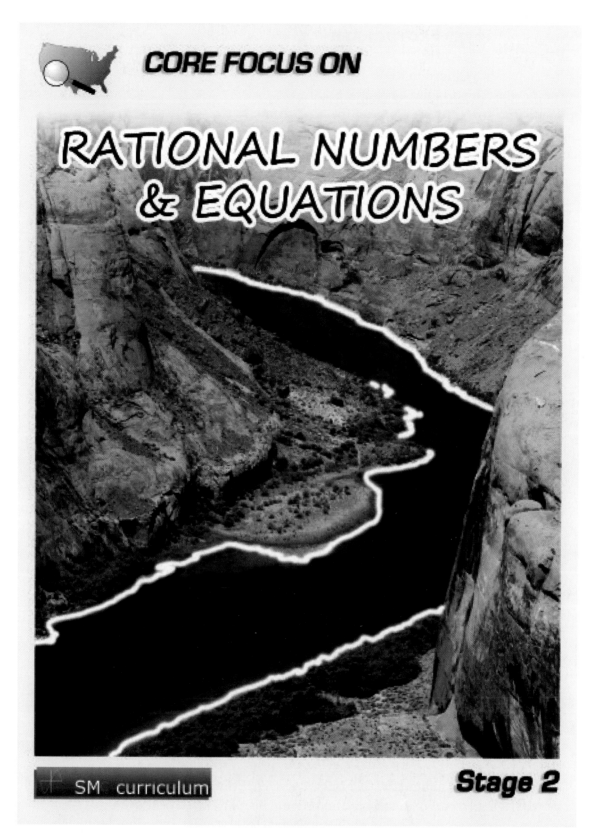

CORE FOCUS ON
RATIONAL NUMBERS & EQUATIONS

Stage 2

AUTHORS

Shannon McCaw

Beth Armstrong • Matt McCaw • Sarah Schuhl • Michelle Terry • Scott Valway

COVER PHOTOGRAPH

Colorado River

One of the most important rivers in the Southwest, the Colorado River also provides some of the most scenic whitewater rafting in the world. Seen here at Horseshoe Bend, the Colorado River carved the Grand Canyon, a natural wonder of the United States.

©iStockphoto.com/Aneese

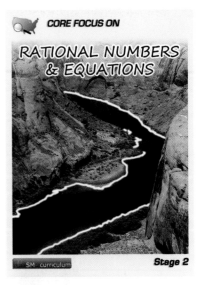

Copyright ©2014 by SM^c Curriculum, LLC. All rights reserved. Printed in the United States. This publication is protected by copyright. No part of this publication should be reproduced or transmitted in any form or by any means without prior written consent of the publisher. This includes, but is not limited to, electronic reproduction, storage in a retrieval system, photocopying, recording or broadcasting for distance learning. For information regarding permission(s), write to: Permissions Department.

ISBN: 978-1-938801-73-0

2 3 4 5 6 7 8 9 10

ABOUT THE AUTHORS

From left to right: Beth Armstrong, Matt McCaw, Shannon McCaw, Scott Valway, Michelle Terry, Sarah Schuhl

SERIES AUTHOR

Shannon McCaw has been a classroom teacher in the Newberg and Parkrose School Districts in Oregon. She has been trained in Professional Learning Communities, Differentiated Instruction and Critical Friends. Shannon currently works as a consultant with math teachers from over 100 districts around Oregon. Shannon's areas of expertise include the Common Core State Standards, curriculum alignment, assessment best practices and instructional strategies. She has a degree in Mathematics from George Fox University and a Masters of Arts in Secondary Math Education from Colorado College.

CONTRIBUTING AUTHORS & EDITORS

Beth Armstrong has been a classroom teacher in the Beaverton School District in Oregon. She has received training in Talented and Gifted Instruction. She has a Masters in Curriculum and Instruction from Washington State University.

Matt McCaw has been a classroom teacher, math/science TOSA and special education case-manager in several Oregon school districts. Matt has most recently worked as a curriculum developer and math coach for grades 6-8. He is trained in Differentiated Instruction, Professional Learning Communities, Critical Friends Groups and Understanding Poverty. Matt has a Masters of Special Education from Western Oregon University.

Sarah Schuhl has been a classroom teacher, secondary math instructional coach and district-wide K-12 math instructional specialist, most recently in the Centennial School District in Oregon. Sarah currently works as a Solution Tree associate and an educational consultant supporting and challenging teachers in the areas of math instruction and alignment to the Common Core State Standards, common assessments for all subjects and grade levels and professional learning communities. From 2010–2013, Sarah served as a member and chair of the National Council of Teachers of Mathematics editorial panel for their Mathematics Teacher journal. Sarah earned a Masters of Science in Teaching Mathematics from Portland State University.

Michelle Terry has been a classroom teacher in the Estacada and Newberg School Districts in Oregon. Michelle has received training in Professional Learning Communities, Critical Friends, ELL Instructional Strategies, Proficiency-Based Grading and Lesson Design, Power Strategies for Effective Teaching, and Classroom Love and Logic. Michelle has an Interdisciplinary Masters from Western Oregon University. She currently teaches mathematics at Newberg High School.

Scott Valway has been a classroom teacher in the Tigard-Tualatin, Newberg and Parkrose School Districts in Oregon. Scott has been trained in Differentiated Instruction, Professional Learning Communities, Critical Friends, Discovering Algebra, Pre-Advanced Placement, Assessment Writing and Credit by Proficiency. Scott has a Masters of Science in Teaching from Oregon State University. He currently teaches math at Parkrose High School.

COMMON CORE STATE STANDARDS

Grade 7 Overview

The complete set of Common Core State Standards can be found at http://www.corestandards.org/. This book focuses on the highlighted Common Core State Standards shown below.

Ratios and Proportional Relationships

- Analyze proportional relationships and use them to solve real-world and mathematical problems.

The Number System

- Apply and extend previous understanding of operations with fractions to add, subtract, multiply and divide rational numbers.

Expressions and Equations

- Use properties of operations to generate equivalent expressions.

- Solve real-life and mathematical problems using numerical and algebraic expressions and equations.

Geometry

- Draw, construct and describe geometrical figures and describe the relationships between them.

- Solve real-life and mathematical problems involving angle measure, area, surface area and volume.

Statistics and Probability

- Use random sampling to draw inferences about a population.

- Draw informal comparative inferences about two populations.

- Investigate chance processes and develop, use and evaluate probability models.

Mathematical Practices

1. Make sense of problems and persevere in solving them.

2. Reason abstractly and quantitatively.

3. Construct viable arguments and critique the reasoning of others.

4. Model with mathematics.

5. Use appropriate tools strategically.

6. Attend to precision.

7. Look for and make use of structure.

8. Look for and express regularity in repeated reasoning.

CORE FOCUS ON RATIONAL NUMBERS & EQUATIONS

CONTENTS IN BRIEF

How To Use Your Math Book	VIII
Block 1 Positive Ratonal Numbers	1
Block 2 Integers	37
Block 3 Rational Number Operations	73
Block 4 Solving Equations	106
Acknowledgements	147
English/Spanish Glossary	148
Selected Answers	183
Index	186
Problem-Solving	188
Symbols	189

CORE FOCUS ON RATIONAL NUMBERS & EQUATIONS

BLOCK 1 ~ POSITIVE RATIONAL NUMBERS

Lesson 1.1	Simplifying Fractions	3
	Explore! Fraction Tiles	
Lesson 1.2	Mixed Numbers and Improper Fractions	8
Lesson 1.3	Adding & Subtracting Fractions	11
	Explore! Fraction Careers	
Lesson 1.4	Multiplying and Dividing Fractions	15
Lesson 1.5	Operations with Mixed Numbers	20
	Explore! Rope Rodeo	
Lesson 1.6	Adding and Subtracting Decimals	25
Lesson 1.7	Multiplying and Dividing Decimals	28
Review	Block 1 ~ Positive Rational Numbers	32

BLOCK 2 ~ INTEGERS

Lesson 2.1	Understanding Integers	39
Lesson 2.2	Adding Integers	44
	Explore! Integer Chips	
Lesson 2.3	Subtracting Integers	49
Lesson 2.4	Multiplying Integers	52
	Explore! Number Jumping	
Lesson 2.5	Dividing Integers	56
Lesson 2.6	Powers and Exponents	60
	Explore! Positive or Negative?	
Lesson 2.7	Order of Operations	64
	Explore! Fact Puzzle	
Review	Block 2 ~ Integers	68

BLOCK 3 ~ RATIONAL NUMBER OPERATIONS

Lesson 3.1	Estimating Sums and Differences	75
	Explore! Trip to the Store	
Lesson 3.2	Adding Rational Numbers	80
Lesson 3.3	Subtracting Rational Numbers	84
	Explore! What's the Difference?	
Lesson 3.4	Estimating Products and Quotients	88
	Explore! In Your Head	
Lesson 3.5	Multiplying Rational Numbers	92
Lesson 3.6	Dividing Rational Numbers	97
Review	Block 3 ~ Rational Number Operations	101

BLOCK 4 ~ SOLVING EQUATIONS

Lesson 4.1	Expressions and Equations	108
Lesson 4.2	Solving One-Step Equations	113
	Explore! Introduction to Equation Mats	
Lesson 4.3	Solving Two-Step Equations	118
	Explore! Equation Mats for Two-Step Equations	
Lesson 4.4	The Distributive Property	122
Lesson 4.5	Simplifying Expressions	126
	Explore! Where Do I Belong?	
Lesson 4.6	Simplifying and Solving Equations	130
	Explore! Equation Manipulation	
Lesson 4.7	Solving Equations with Variables on Both Sides	134
Lesson 4.8	Linear Inequalities	138
Review	Block 4 ~ Solving Equations	142

HOW TO USE YOUR MATH BOOK

Your math book has features that will help you be successful in this course. Use this guide to help you understand how to use this book.

LESSON TARGET

 Look in this box at the beginning of every lesson to know what you will be learning about in each lesson.

VOCABULARY

Each new vocabulary word is printed in **red**. The definition can be found with the word. You can also find the definition of the word in the glossary which is in the back of this book.

EXPLORE!

Some lessons have **EXPLORE!** activities which allow you to discover mathematical concepts. Look for these activities in the Table of Contents and in lessons next to the purple line.

EXAMPLES

Examples are useful because they remind you how to work through different types of problems. Look for the word **EXAMPLE** and the green line.

HELPFUL HINTS

Helpful hints and important things to remember can be found in green callout boxes.

BLUE BOXES

A blue box holds important information or a process that will be used in that lesson. Not every lesson has a blue box.

 This calculator icon will appear in Lessons and Exercises where a calculator is needed. Your teacher may want you to use your calculator at other times, too. If you are unsure, make sure to ask if it is the right time to use it.

EXERCISES

The **EXERCISES** are a place for you to find practice problems to determine if you understand the lesson's target. You can find selected answers in the back of this book so you can check your progress.

REVIEW

The **REVIEW** provides a set of problems for you to practice concepts you have already learned in this book. The **REVIEW** follows the **EXERCISES** in each lesson. There is also a **REVIEW** section at the end of each Block.

TIC-TAC-TOE ACTIVITIES

Each Block has a Tic-Tac-Toe board at the beginning with activities that extend beyond the Common Core State Standards. The Tic-Tac-Toe activities described on the board can be found throughout each Block in yellow boxes.

CAREER FOCUS

At the end of each Block, you will find an autobiography of an individual. Each one describes what they like about their job and how math is used in their career.

CORE FOCUS ON MATH
STAGE 2

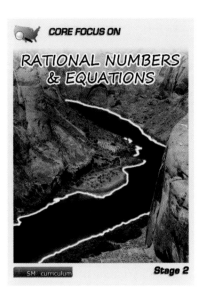

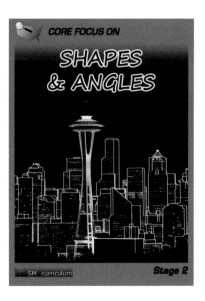

CORE FOCUS ON RATIONAL NUMBERS & EQUATIONS

BLOCK 1 ~ POSITIVE RATIONAL NUMBERS

LESSON 1.1	SIMPLIFYING FRACTIONS	3
	EXPLORE! FRACTION TILES	
LESSON 1.2	MIXED NUMBERS AND IMPROPER FRACTIONS	8
LESSON 1.3	ADDING & SUBTRACTING FRACTIONS	11
	EXPLORE! FRACTION CAREERS	
LESSON 1.4	MULTIPLYING AND DIVIDING FRACTIONS	15
LESSON 1.5	OPERATIONS WITH MIXED NUMBERS	20
	EXPLORE! ROPE RODEO	
LESSON 1.6	ADDING AND SUBTRACTING DECIMALS	25
LESSON 1.7	MULTIPLYING AND DIVIDING DECIMALS	28
REVIEW	BLOCK 1 ~ POSITIVE RATIONAL NUMBERS	32

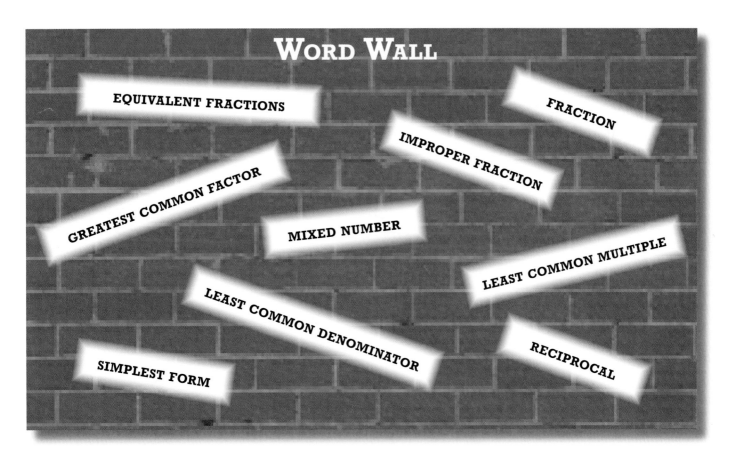

BLOCK 1 ~ POSITIVE RATIONAL NUMBERS

TIC-TAC-TOE

WHICH IS THE BIGGEST? Order and compare fractions with unlike denominators. *See page 7 for details.*	**REAL-WORLD USES** Survey people about how they use fractions and decimals in their everyday lives. Write a report summarizing your findings. *See page 19 for details.*	**GCF AND LCM** Write two different poems about the greatest common factor and least common multiple. *See page 19 for details.*
MEMORY GAME Create a matching game using fraction and decimal expressions and their solutions. *See page 31 for details.*	**NUMBERS GALORE** Find the value of fraction or decimal expressions including more than two numbers. *See page 31 for details.*	**RECIPE CONVERSION** Write one recipe in 5 different formats. Give situations where each format would be useful. *See page 10 for details.*
DECIMALS OR FRACTIONS? Write a play where the characters, decimals and fractions, are each trying to prove they are the easiest to work with. *See page 35 for details.*	**GRID GAME** Create a grid game using decimal expressions that include the four operations (+, −, ×, ÷). *See page 24 for details.*	**HARDWARE STORE** Take a trip to a hardware store to find at least 10 items described by fractions. Create a display of the items. *See page 7 for details.*

Block 1 ~ Positive Rational Numbers ~ Tic-Tac-Toe

SIMPLIFYING FRACTIONS

LESSON 1.1

 Write fractions in simplest form.

Karl and Kevin ordered a pizza for lunch. The large pepperoni pizza came with 8 evenly cut slices. Karl ate two pieces and Kevin ate $\frac{1}{4}$ of the pizza. Who ate more?

A **fraction** is a number that represents a part of a whole number. Fractions represent division of two numbers or a comparison of two numbers. A fraction is written in the form $\frac{\text{numerator}}{\text{denominator}}$. In a fraction, the denominator cannot be 0.

In the pizza problem, you can compare 2 slices to the entire pizza of 8 slices as $\frac{2}{8}$.

To figure out who ate more pizza, determine whether $\frac{2}{8}$ or $\frac{1}{4}$ is larger. Look at the diagrams below. It turns out that both boys ate equal portions of the pizza. One quarter $\left(\frac{1}{4}\right)$ of the pizza is the same as 2 out of 8 slices $\left(\frac{2}{8}\right)$. These two fractions are called equivalent fractions. **Equivalent fractions** are fractions that represent the same number but have different numerators and denominators.

Karl's Portion  =  Kevin's Portion

Equivalent fractions can be found by multiplying or dividing the numerator and denominator by the same non-zero number. If you are dividing both the numerator and denominator to find an equivalent fraction, the number must be a factor other than 1 that is common to both numbers.

EXAMPLE 1   Find two fractions that are equivalent to $\frac{10}{14}$.

SOLUTION   There are an infinite number of equivalent fractions. When multiplying the top and bottom of the fraction to create an equivalent fraction, you may choose any number.

To find an equivalent fraction by dividing the numerator and denominator, the number you are dividing by must be a common factor. In this case, the only common factor of 10 and 14 is 2.

Lesson 1.1 ~ Simplifying Fractions   **3**

EXPLORE! FRACTION TILES

Each large rectangle below represents one whole. Each rectangle is divided into equal parts. It is called a fraction bar.

Step 1: Write a fraction to represent $\frac{\text{number of shaded sections}}{\text{total number of sections}}$ for each fraction bar.

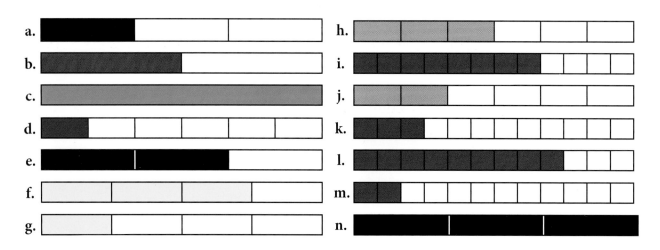

Step 2: Match each fraction in the left column with its *equivalent* fraction in the right column. Write the pair of equivalent fractions. For example, $\frac{1}{5} = \frac{2}{10}$.

Step 3: Write a third equivalent fraction for each pair in **Step 2**.

Step 4: Draw two fraction bars to show $\frac{3}{4}$ and $\frac{6}{8}$. Are these equivalent fractions? Explain. If so, write a third equivalent fraction to $\frac{3}{4}$ and $\frac{6}{8}$.

Step 5: The fractions in the left column are in *simplest form*. Look at the fractions in the left column compared to those in the right column and explain what *simplest form* means.

The **greatest common factor** (GCF) is the greatest factor that is common to two or more numbers. To determine the greatest common factor, find the prime factorization of each number. The GCF is the product of the prime factors which are common to both numbers.

EXAMPLE 2   **Find the GCF of 24 and 36.**

SOLUTION   Look at the prime factorization of 24 and 36.

$$24 = ②\cdot ②\cdot 2 \cdot ③ \qquad 36 = ②\cdot ②\cdot ③\cdot 3$$

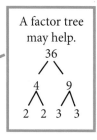

A factor tree may help.

The common prime factors of 24 and 36 are 2, 2 and 3. The GCF can be found by multiplying these factors together.

$$2 \cdot 2 \cdot 3 = 12$$
$$\text{GCF} = 12$$

A fraction is in **simplest form** when the numerator and the denominator's only common factor is 1. When you divide the numerator and denominator by its greatest common factor, the fraction will be in its simplest form. You can also get a fraction into simplest form by repeatedly dividing by common factors until the only common factor between the numerator and denominator is 1.

> **WRITING FRACTIONS IN SIMPLEST FORM**
> Divide the numerator and denominator by the greatest common factor (GCF).
> **OR**
> Divide the numerator and denominator by common factors until the only common factor is 1.

EXAMPLE 3   Write each fraction in simplest form.

a. $\frac{20}{50}$ b. $\frac{45}{60}$

SOLUTIONS

a. Find the GCF of 20 and 50.

$20 = 2 \cdot ②\cdot ⑤ \qquad 50 = ②\cdot ⑤\cdot 5$

The common prime factors of 20 and 50 are 2 and 5.
GCF = 2 · 5 = 10

Divide the numerator and denominator by the GCF.

$\frac{20}{50}$ in simplest form is $\frac{2}{5}$

b. One common factor of 45 and 60 is 5.
Divide the numerator and denominator by 5.

A common factor of 9 and 12 is 3. Divide the numerator and denominator by 3.

$\frac{45}{60}$ in simplest form is $\frac{3}{4}$

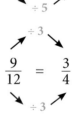

Lesson 1.1 ~ Simplifying Fractions

EXERCISES

Write two fractions that are equivalent to the given fraction.

1. $\frac{8}{16}$

2. $\frac{5}{15}$

3. $\frac{4}{10}$

4. $\frac{4}{14}$

5. $\frac{2}{8}$

6. $\frac{6}{9}$

Find the greatest common factor (GCF) of each pair of numbers.

7. 4 and 8

8. 10 and 15

9. 12 and 30

10. 30 and 40

11. 40 and 60

12. 6 and 15

Write each fraction in simplest form.

13. $\frac{12}{18}$

14. $\frac{20}{50}$

15. $\frac{4}{12}$

16. $\frac{10}{24}$

17. $\frac{6}{14}$

18. $\frac{24}{42}$

19. $\frac{6}{15}$

20. $\frac{16}{30}$

21. $\frac{18}{27}$

22. $\frac{75}{100}$

23. $\frac{20}{35}$

24. $\frac{120}{150}$

Determine if each pair of fractions is equivalent. Use words and/or numbers to show how you determined your answer.

25. $\frac{6}{8}$ and $\frac{9}{12}$

26. $\frac{10}{20}$ and $\frac{14}{24}$

27. $\frac{20}{30}$ and $\frac{16}{24}$

28. Fractions can be simplified by dividing the numerator and denominator by the GCF or by dividing the numerator and denominator by a common factor until the only common factor that remains between the numerator and denominator is 1. Which method do you prefer to use? Why?

29. A worm measured $\frac{6}{16}$ foot long using a ruler. Simplify this measurement.

30. Carmen weighed her baby kitten and found that he weighed $\frac{28}{32}$ pound. Simplify this measurement.

31. Write two equivalent fractions to represent the following fraction bar:

32. A student claims that $\frac{26}{39}$ is in simplest form. Do you agree? Explain.

Tic-Tac-Toe ~ Which Is The Biggest?

In this activity you will arrange sets of fractions from least to greatest. In order to compare fractions, it helps if the fractions have a common denominator. For each set of fractions, find the least common denominator (LCD) and write the equivalent fractions using the LCD. Once the fractions have a common denominator, the numerators can be compared. When you write the set of numbers from least to greatest, you must use the original fractions in simplest form.

For example: List the following fractions from least to greatest. $\frac{2}{5}, \frac{1}{4}, \frac{3}{10}$

Rewrite all fraction using the LCD of 20.

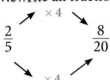

$\frac{2}{5} \rightarrow \frac{8}{20}$ $\frac{1}{4} \rightarrow \frac{5}{20}$ $\frac{3}{10} \rightarrow \frac{6}{20}$

List the fractions from least to greatest. $\frac{5}{20}, \frac{6}{20}, \frac{8}{20}$

Rewrite the fractions in simplest form. $\frac{1}{4}, \frac{3}{10}, \frac{2}{5}$

List each set of fractions from least to greatest.

1. $\frac{1}{2}, \frac{3}{4}, \frac{5}{8}$

2. $\frac{7}{15}, \frac{2}{5}, \frac{7}{10}$

3. $\frac{5}{6}, \frac{11}{12}, \frac{23}{24}$

4. $\frac{23}{24}, \frac{1}{6}, \frac{2}{9}, \frac{5}{12}$

5. $\frac{1}{2}, \frac{3}{7}, \frac{5}{14}, \frac{4}{7}$

6. $\frac{3}{5}, \frac{13}{20}, \frac{5}{8}, \frac{23}{40}$

7. Make your own list of four fractions with different denominators and list them from least to greatest.

Tic-Tac-Toe ~ Hardware Store

Take a trip to a local hardware store. Find at least 10 items that are described using fractions or mixed numbers. Make sure you have a wide variety of items. Create a display of the items by either sketching the item or using pictures. Include a table, as seen below, in your display.

Example:

Picture	Item	Fraction Use	Possible Item Use
	Acorn Nut Sleeve Anchor	Size: $\frac{1}{4} \times 1\frac{3}{8}$ in	Tightening objects

Lesson 1.1 ~ Simplifying Fractions

MIXED NUMBERS AND IMPROPER FRACTIONS

LESSON 1.2

Write mixed numbers as improper fractions.
Write improper fractions as mixed numbers.

Bali measured the length of her pointer finger with a ruler. It was $2\frac{3}{4}$ inches long. This fraction is called a **mixed number**. A mixed number is the sum of a whole number and a fraction less than 1.

Mixed numbers can also be written as **improper fractions**. An improper fraction has a numerator that is greater than or equal to its denominator.

Bali's finger was $2\frac{3}{4}$ inches long. In order to write this fraction as an improper fraction, she must determine how many "fourths" are in $2\frac{3}{4}$.

 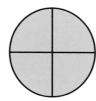 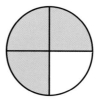

Bali drew out the three circles above and shaded in $2\frac{3}{4}$ of the circles. She is able to see that there are 11 "fourths" which are shaded. The improper fraction for $2\frac{3}{4}$ is $\frac{11}{4}$ which is read "eleven fourths". Mixed numbers can be rewritten as improper fractions by following a mathematical process involving multiplication and addition.

REWRITING MIXED NUMBERS AS IMPROPER FRACTIONS

1. Multiply the whole number by the denominator.
2. Add the numerator to the product.
3. Write this number as the numerator and keep the original denominator as the denominator of the improper fraction.

EXAMPLE 1   Write $5\frac{1}{6}$ as an improper fraction.

SOLUTION

Multiply the whole number by the denominator.	$5(6) = 30$
Add the numerator to the product.	$1 + 30 = 31$
Write the improper fraction as the total over the original denominator.	$\frac{31}{6}$

8   Lesson 1.2 ~ Mixed Numbers and Improper Fractions

Improper fractions can also be rewritten as mixed numbers. In most situations, solutions are considered in simplest form when written as a mixed number.

REWRITING IMPROPER FRACTIONS AS MIXED NUMBERS

1. Divide the numerator by the denominator. The quotient is the whole number in the mixed number.
2. Write the remainder as the numerator over the original denominator. This is the fraction in the mixed number.

EXAMPLE 2   Write $\frac{17}{5}$ as a mixed number.

SOLUTION   Divide the numerator by the denominator.

$$5\overline{\smash{)}17}$$ quotient 3, remainder 2

Whole number = 3

Put the remainder over the original denominator. Fraction = $\frac{2}{5}$

Write the mixed number with both the whole number and the fraction. $3\frac{2}{5}$

EXERCISES

Write the improper fraction and mixed number for each drawing.

1.

2.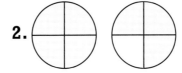

3.

4.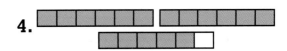

Write each mixed number as an improper fraction.

5. $2\frac{1}{2}$

6. $3\frac{2}{5}$

7. $1\frac{6}{7}$

8. $6\frac{1}{3}$

9. $4\frac{3}{4}$

10. $5\frac{7}{10}$

11. $1\frac{4}{9}$

12. $7\frac{7}{8}$

13. $8\frac{2}{3}$

14. Jorge's mom bought a house in Memphis. Her interest rate on her loan was $5\frac{3}{4}\%$. Write her interest rate as an improper fraction.

15. At the district track and field meet, two girls tied for first in the high jump. Both girls cleared $5\frac{5}{12}$ feet. Write this height as an improper fraction.

16. A chef is making chicken parmesan for a banquet. His recipe calls for $3\frac{3}{8}$ cups of tomato sauce. He put in $\frac{25}{8}$ cups. Did he put in enough? Show all work necessary to justify your answer.

Write each improper fraction as a mixed number.

17. $\frac{8}{5}$

18. $\frac{10}{3}$

19. $\frac{11}{4}$

20. $\frac{21}{8}$

21. $\frac{43}{10}$

22. $\frac{29}{9}$

23. $\frac{13}{2}$

24. $\frac{44}{5}$

25. $\frac{15}{4}$

26. There are several thousand species of beetles. The body of some beetles could be as long as $\frac{5}{4}$ inches. Write this length as a mixed number.

27. Mike's hair has grown $\frac{31}{10}$ centimeters since his last hair cut. Sarah's hair has grown 3 centimeters. Who's hair has grown more? Explain your reasoning.

28. Kaitlyn visited with her grandma for 81 minutes. Write this amount as a mixed number that represents the number of hours she visited with her grandma.

REVIEW

Write each fraction in simplest form.

29. $\frac{10}{35}$

30. $\frac{25}{100}$

31. $\frac{8}{24}$

32. $\frac{12}{16}$

33. $\frac{14}{20}$

34. $\frac{22}{30}$

Tic-Tac-Toe ~ Recipe Conversion

Find a recipe with a minimum of 8 ingredients. Create a recipe booklet with the recipe written five different ways. For example, you could cut the amount of each ingredient in half or triple the recipe. You could also make all ingredients have a common denominator. On each page, explain how the recipe was converted. Also, give situations where each version of the recipe might be useful. Be creative with your recipe booklet. Include illustrations and an interesting cover for the booklet.

For example: Double the Recipe – Use when you have friends over…
Cut the Recipe in Fourths – Use when cooking for yourself…

ADDING AND SUBTRACTING FRACTIONS

LESSON 1.3

 Add and subtract fractions with like and unlike denominators.

Carpenters, chefs, engineers, surveyors and architects add and subtract fractions often as a part of their work. It is important to remember that two fractions must have a common denominator before you can find their sum or difference. Common denominators can be found by using the least common multiple of the two denominators. The **least common multiple (LCM)** is the smallest non-zero multiple that is common to two or more numbers.

EXAMPLE 1 **Find the least common multiple of:**
a. 8 and 12   b. 4 and 5

SOLUTIONS a. One method for finding the least common multiple is to list out the multiples of each number until the same number appears on each list.
8, 16, 24, 32, 40 …
12, 24, 36, 48, 60, 72 …
The LCM is 24.

b. Another method to find the LCM is to choose the largest number, in this case 5, and look at its multiples.
5, 10, 15, 20, 25, 30, 35 …

Find the smallest multiple that 4 divides into evenly.
5 – NO
10 – NO
15 – NO
20 – YES (20 ÷ 4 = 5)
The LCM is 20.

When the least common multiple of two denominators is found, it is called the **least common denominator (LCD)**. Based on the answer in **Example 1**, the LCD of $\frac{1}{8}$ and $\frac{5}{12}$ is 24 because the least common multiple of 8 and 12 is 24.

ADDING AND SUBTRACTING FRACTIONS

1. If denominators are not equal, rewrite the fractions as equivalent fractions with common denominators using the least common denominator (LCD).
2. Add or subtract the numerators.
3. Write the sum or difference over the common denominator.
4. Write the fraction in simplest form. If the sum or difference is an improper fraction it should be changed to a mixed number.

Lesson 1.3 ~ Adding and Subtracting Fractions **11**

EXAMPLE 2 Find the value of $\frac{1}{10} + \frac{7}{10}$.

SOLUTION These fractions already have a common denominator. Add the numerators.

$$\frac{1+7}{10} = \frac{8}{10}$$

Write in simplest form.

$$\frac{8}{10} \overset{\div 2}{\underset{\div 2}{=}} \frac{4}{5}$$

$$\frac{1}{10} + \frac{7}{10} = \frac{4}{5}$$

EXAMPLE 3 On Monday, the New Orleans International Airport reported $\frac{3}{8}$ inch of rain. On Tuesday it rained $\frac{11}{16}$ inch. What was the total rainfall for these two days?

SOLUTION The least common multiple of 8 and 16 is 16. Write $\frac{3}{8}$ as an equivalent fraction with a denominator of 16.

$$\frac{3}{8} \overset{\times 2}{\underset{\times 2}{=}} \frac{6}{16}$$

Write the sum over the LCD.

$$\frac{6}{16} + \frac{11}{16} = \frac{17}{16}$$

Change from an improper fraction to a mixed number.

$$\frac{17}{16} = 1\frac{1}{16}$$

It rained a total of $1\frac{1}{16}$ inches in New Orleans on Monday and Tuesday.

EXAMPLE 4 Find the value of $\frac{2}{3} - \frac{1}{4}$.

SOLUTION The least common multiple of 3 and 4 is 12. Write each fraction as equivalent fractions with a denominator of 12.

$$\frac{2}{3} \overset{\times 4}{\underset{\times 4}{=}} \frac{8}{12} \qquad \frac{1}{4} \overset{\times 3}{\underset{\times 3}{=}} \frac{3}{12}$$

Write the difference over the LCD.

$$\frac{8}{12} - \frac{3}{12} = \frac{5}{12}$$

$$\frac{2}{3} - \frac{1}{4} = \frac{5}{12}$$

Lesson 1.3 ~ Adding and Subtracting Fractions

EXPLORE! — FRACTION CAREERS

In each step you will perform a calculation that might be done in a certain career. Determine the solution to each step and then guess in what career you might do that specific calculation.

Step 1: Kara mixed $\frac{2}{5}$ gallon of red paint with $\frac{7}{10}$ gallon of blue paint to create a deep indigo paint. How much total indigo paint did Kara create? Name a possible career that Kara may have.

Step 2: Jillian made 36 pastries for a Mother's Day brunch. Before making the pastries, she had $\frac{7}{8}$ cup butter. After making the pastries, only $\frac{1}{16}$ cup of butter remained. How much butter was used for the pastries? What career might Jillian have?

Step 3: Rick nailed two boards together. One board was $\frac{3}{8}$ inch thick. The other board was $\frac{1}{2}$ inch thick. What is the total thickness of the two boards nailed together? What is a possible career Rick might have?

Step 4: Hans helped a family get a home loan for their first house. He told the family there was a $\frac{1}{4}$% fee for the title company and another $\frac{3}{4}$% fee for the listing agent. What was the total percentage of fees the family will have to pay? What do you think Hans does for a living?

Step 5: Jocelyn had $\frac{19}{20}$ liter of a cleaning solution in a container. She used $\frac{1}{10}$ liter of the solution on a stain in the carpet. How many liters of solution remain in the container? What do you think Jocelyn's career is?

Step 6: Name another career that uses fractions. Make up a story problem that would involve adding or subtracting fractions in that career. Find the answer to your problem.

EXERCISES

Find the least common multiple (LCM) of the numbers given.

1. 3 and 6
2. 3 and 5
3. 4 and 9
4. 10 and 15
5. 12 and 20
6. 9 and 15

7. Jaime is convinced that his answer for the problem below is correct. Look at his work below. Determine if Jaime's answer is correct. If it is not correct, explain why the answer is incorrect and find the correct answer.

$$\frac{1}{4} + \frac{3}{8} = \frac{4}{12} \longrightarrow \frac{4}{12} = \boxed{\frac{1}{3}}$$

Find each sum or difference. Write your answer in simplest form.

8. $\frac{2}{7} + \frac{3}{7}$
9. $\frac{9}{10} - \frac{3}{10}$
10. $\frac{1}{3} + \frac{1}{6}$

Lesson 1.3 ~ Adding and Subtracting Fractions

11. $\frac{7}{8} - \frac{3}{16}$

12. $\frac{5}{6} - \frac{1}{3}$

13. $\frac{2}{3} + \frac{1}{2}$

14. $\frac{2}{3} - \frac{1}{6}$

15. $\frac{2}{15} + \frac{3}{10}$

16. $\frac{37}{40} + \frac{2}{5}$

17. $\frac{11}{12} + \frac{1}{2}$

18. $\frac{3}{5} - \frac{1}{4}$

19. $\frac{7}{10} + \frac{3}{4}$

20. One day after school Keenan worked on his math homework for $\frac{4}{5}$ hour and his English homework for $\frac{1}{4}$ hour. How long did Keenan work on homework for those two subjects? Show all work necessary to justify your answer.

21. A carpenter measured the thickness of two pieces of plywood. One board was $\frac{13}{16}$ inch thick. The other board was $\frac{5}{8}$ inch thick. How much thicker is one board than the other?

22. A tomato plant measured $\frac{3}{10}$ meter when it was planted. Two weeks later the plant measured $\frac{17}{20}$ meter. How much did the plant grow in those two weeks?

23. A cookie recipe called for $\frac{3}{4}$ cup of flour at the beginning and another $\frac{2}{3}$ cup later in the recipe. What is the total amount of flour needed to make the cookies?

24. Five students sold raffle tickets at a school fundraising auction. The table at right shows the portion of the total sales for each of the five students.

Student	Fraction of Total Raffle Ticket Sales
Patrick	$\frac{1}{10}$
Clara	$\frac{7}{20}$
Maria	$\frac{1}{4}$
Lamar	$\frac{1}{12}$
Lisa	$\frac{13}{60}$

a. Clara and Maria were the top sellers. What fraction of the raffle tickets did they sell?

b. What is the difference between Lisa's and Lamar's fraction of raffle ticket sales?

25. Ian's family was traveling from Denver, Colorado to Milwaukee, Wisconsin to visit relatives. On the first day they went $\frac{1}{3}$ of the total distance. On the second day of the trip, they went $\frac{2}{5}$ of the total distance. What fraction of the trip had they traveled after two days? Explain how you know your answer is correct.

REVIEW

Write each mixed number as an improper fraction.

26. $3\frac{3}{7}$

27. $2\frac{1}{5}$

28. $8\frac{2}{3}$

Write each improper fraction as a mixed number.

29. $\frac{11}{2}$

30. $\frac{14}{5}$

31. $\frac{59}{10}$

MULTIPLYING AND DIVIDING FRACTIONS

LESSON 1.4

 Multiply and divide fractions.

Caroline was making waffles which required $\frac{3}{4}$ cup milk. She wanted to divide the recipe in half because she was only making waffles for two people. How could she find $\frac{1}{2}$ of $\frac{3}{4}$?

In math, the word "of" often means multiply. In this case, Caroline can find $\frac{1}{2}$ of $\frac{3}{4}$ by multiplying $\frac{1}{2} \cdot \frac{3}{4}$. To see a model of this, use a fraction bar:

The shaded part represents $\frac{3}{4}$.

To find $\frac{1}{2}$ of $\frac{3}{4}$, divide the rectangle horizontally in half. Shade one of the two horizontal regions with a different color to represent $\frac{1}{2}$.

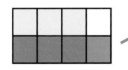

Shade either the top or bottom row blue.

Count the number of sections that have been shaded twice (where yellow and blue overlap to make green). Three sections out of eight are shaded green. This means that $\frac{1}{2}$ of $\frac{3}{4}$ is $\frac{3}{8}$. Caroline will need $\frac{3}{8}$ cup of milk to make half the recipe of waffles.

According to the fraction model above, $\frac{1}{2} \cdot \frac{3}{4} = \frac{3}{8}$. When you find the product of the numerators, you will have the numerator of the answer. When you multiply the denominators together, you will have the denominator of the answer. This is true for all multiplication problems involving fractions. Some fractions will need to be simplified after multiplying the numerator and denominator.

MULTIPLYING FRACTIONS

For any numbers a, b, c and d:
$$\frac{a}{b} \cdot \frac{c}{d} = \frac{a \cdot c}{b \cdot d}$$

EXAMPLE 1 Find the value of $\frac{2}{3} \cdot \frac{3}{5}$ using models.

SOLUTION Draw a model and shade $\frac{3}{5}$ of a rectangle. Draw horizontal lines to cut the model into thirds. Shade 2 of the three horizontal rectangles to represent $\frac{2}{3}$.

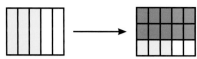

The final model has 6 out of 15 regions which are double shaded, so $\frac{2}{3} \cdot \frac{3}{5} = \frac{6}{15}$.

In simplest form, $\frac{6}{15} = \frac{2}{5}$.

EXAMPLE 2 Find the value of $\frac{5}{6} \cdot \frac{2}{5}$.

SOLUTION Multiply.

Simplify.

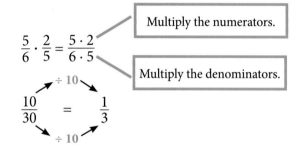

$\frac{5}{6} \cdot \frac{2}{3} = \frac{1}{3}$

Two numbers are **reciprocals** if their product is 1. To find the reciprocal of a fraction, "flip" the fraction so that the numerator becomes the denominator and the denominator becomes the numerator.

$\frac{3}{4}$ → Reciprocal → $\frac{4}{3}$ $\frac{1}{2}$ → Reciprocal → $\frac{2}{1}$

Reciprocals are used when dividing fractions. To divide by a fraction, multiply by its reciprocal.

DIVIDING FRACTIONS

For any numbers *a*, *b*, *c* and *d*:
$$\frac{a}{b} \div \frac{c}{d} = \frac{a \cdot d}{b \cdot c}$$

Lesson 1.4 ~ Multiplying and Dividing Fractions

EXAMPLE 3 Find the value of $\frac{3}{5} \div \frac{7}{10}$.

SOLUTION

Divide.

$$\frac{3}{5} \div \frac{7}{10} = \frac{3}{5} \cdot \frac{10}{7} = \frac{30}{35}$$

Simplify.

$$\frac{30}{35} \xrightarrow{\div 5} = \xrightarrow{\div 5} \frac{6}{7}$$

$$\frac{3}{5} \div \frac{7}{10} = \frac{6}{7}$$

Multiply the first fraction by the reciprocal of the second fraction.

EXAMPLE 4 Find the value of $\frac{5}{6} \div \frac{1}{8}$.

SOLUTION

Divide.

$$\frac{5}{6} \div \frac{1}{8} = \frac{5}{6} \cdot \frac{8}{1} = \frac{40}{6}$$

Simplify.

$$\frac{40}{6} \xrightarrow{\div 2} = \xrightarrow{\div 2} \frac{20}{3}$$

Change into a mixed number.

$$3 \overline{)20} \\ \underline{-18} \\ 2$$

(quotient 6)

$$\frac{5}{6} \div \frac{1}{8} = 6\frac{2}{3}$$

The remainder goes over the divisor after the whole number.

EXERCISES

1. Find $\frac{1}{2} \cdot \frac{4}{5}$ using fraction bars.
   a. Draw a rectangle and divide it into 5 equal regions by drawing vertical lines.
   b. Shade 4 of the 5 regions to represent $\frac{4}{5}$.
   c. To find $\frac{1}{2}$ of $\frac{4}{5}$, divide the 5 regions in half by drawing a horizontal line through the rectangle.
   d. Shade one of the two horizontal regions with a different color.
   e. Write a fraction representing the number of regions that are shaded twice out of the total number of regions.
   f. $\frac{1}{2} \cdot \frac{4}{5} =$

2. Silvia is making chocolate chip cookies. The recipe calls for $\frac{2}{3}$ cup of sugar. She wants to make $\frac{1}{2}$ batch of cookies. How much sugar will Silvia need? Create a model to support your answer.

Lesson 1.4 ~ Multiplying and Dividing Fractions

Find each product. Write your answer in simplest form.

3. $\frac{3}{4} \cdot \frac{3}{5}$

4. $\frac{2}{3} \cdot \frac{1}{3}$

5. $\frac{4}{7} \cdot \frac{1}{2}$

6. $\frac{5}{8} \cdot \frac{3}{4}$

7. $\frac{1}{3} \cdot \frac{1}{3}$

8. $\frac{4}{7} \cdot \frac{1}{3}$

9. $\frac{3}{4} \cdot \frac{2}{5}$

10. $\frac{7}{8} \cdot \frac{2}{7}$

11. $\frac{1}{3} \cdot \frac{3}{5}$

12. $\frac{4}{5} \cdot \frac{1}{2}$

13. $\frac{10}{5} \cdot \frac{1}{5}$

14. $\frac{9}{10} \cdot \frac{2}{3}$

15. Penelope walks $\frac{7}{10}$ mile to school every day. One day, she walked halfway to school before she remembered she forgot her lunch and had to go home to get it. How close was she to school when she turned around?

16. Xavier bought $\frac{6}{7}$ pound of dog food. Each of his three dogs get $\frac{1}{3}$ of the dog food. What fraction of a pound of dog food does each dog get? Show two ways to find the answer.

17. Jared plays on the Harrisville Special Olympics basketball team. Each player on his team gets to be in the game for half of the playing time. The entire game lasts $\frac{2}{3}$ hour. How many minutes will each player play? Show all work necessary to justify your answer.

Write the reciprocal of each number.

18. $\frac{2}{5}$

19. $\frac{1}{3}$

20. $\frac{4}{9}$

Find each quotient. Write your answer in simplest form.

21. $\frac{1}{4} \div \frac{2}{3}$

22. $\frac{3}{7} \div \frac{4}{5}$

23. $\frac{4}{9} \div \frac{1}{5}$

24. $\frac{7}{8} \div \frac{3}{4}$

25. $\frac{2}{3} \div \frac{1}{4}$

26. $\frac{9}{10} \div \frac{3}{5}$

27. $\frac{4}{7} \div \frac{1}{2}$

28. $\frac{5}{6} \div \frac{5}{9}$

29. $\frac{1}{8} \div \frac{3}{4}$

30. Aaron has $\frac{3}{4}$ hour left to do chores. It takes $\frac{1}{4}$ hour to do each chore. How many chores will he be able to do in the remaining time?

31. Lily buys $\frac{7}{8}$ yard of ribbon. She plans to cut ribbons that are each $\frac{1}{16}$ yard long. How many ribbons will she be able to cut?

32. How many $\frac{1}{8}$ cup scoops will you be able to make from $\frac{3}{4}$ cup of cookie dough? Show how you know your answer is correct.

REVIEW

Find each sum or difference. Write the answer in simplest form.

33. $\frac{3}{4} + \frac{1}{8}$

34. $\frac{7}{10} - \frac{1}{5}$

35. $\frac{2}{3} - \frac{1}{2}$

36. $\frac{4}{5} + \frac{17}{20}$

37. $\frac{1}{2} - \frac{3}{8}$

38. $\frac{1}{4} + \frac{1}{3}$

39. Ryan purchased $\frac{3}{4}$ pound of peanuts and $\frac{7}{16}$ pound of almonds. How many total pounds of nuts did Ryan buy? Use mathematics to justify your answer.

Tic-Tac-Toe ~ GCF and LCM

Poetry is an art form that is composed of carefully chosen words to express a greater depth of meaning. Poetry can be written about many different subjects, including mathematics. The greatest common factor (GCF) and least common multiple (LCM) are key elements needed to find the value of fraction operations in simplest form. Write two different cinquain poems as described below. One cinquain should be about the GCF and the other should be about the LCM.

> **Cinquain**
> Poetry with five lines.
> Line 1 has one word (the title).
> Line 2 has two words that describe the title.
> Line 3 has three words that tell the action.
> Line 4 has four words that express the feeling.
> Line 5 has one word which recalls the title.

Tic-Tac-Toe ~ Real-World Uses

Many adults use fractions and decimals daily at work and at home. Create a survey with a minimum of four questions to gather more information about people's uses of fractions and decimals in their everyday lives. Once your survey is approved by your teacher, give the survey to five different adults. Write a 1–2 page report to summarize your findings.

OPERATIONS WITH MIXED NUMBERS

LESSON 1.5

 Add, subtract, multiply or divide mixed numbers.

Mixed numbers are the sum of a whole number and a fraction less than 1. Mixed numbers are used in many situations. You may need $2\frac{1}{4}$ cups of flour when making cookies. You may run $3\frac{1}{2}$ miles. You might buy $1\frac{1}{8}$ pounds of potatoes. Finding sums, differences, products and quotients with mixed numbers can be done by changing each number into an improper fraction before following the procedures you learned in previous lessons.

OPERATIONS WITH MIXED NUMBERS

1. Write each mixed number as an improper fraction.
2. Follow the procedures for adding, subtracting, multiplying or dividing fractions.
3. Write the answer in simplest form. Convert the answer to a mixed number, if needed.

EXPLORE!   **ROPE RODEO**

Four friends participate in the rodeo and purchased rope to use in different events.

Step 1: Keenan purchased two pieces of rope. One piece of rope is $5\frac{1}{4}$ yards long. The other piece is $4\frac{5}{8}$ yards long. Keenan needs help finding the total length of rope he purchased.
    a. Write each mixed number as an improper fraction.
    b. Determine which operation (add, subtract, multiply or divide) should be used.
    c. Follow the rules for that operation. Do you need a common denominator? Do you need the reciprocal? Look at **Lessons 3 and 4** if you need help remembering the procedures.
    d. Simplify your answer and convert it to a mixed number, if needed.

Step 2: Laura bought a piece of rope that is $13\frac{1}{3}$ yards long. She wants to cut it into equal-sized pieces that are each $2\frac{2}{3}$ yards long. Help her determine how many pieces of rope she will have after cutting it.
    a. Write each mixed number as an improper fraction.
    b. Determine which operation (add, subtract, multiply or divide) should be used.
    c. Follow the rules for that operation.
    d. Simplify your answer and convert it to a mixed number, if needed.

EXPLORE! (CONTINUED)

Step 3: Olivia bought a piece of rope that was $2\frac{1}{5}$ yards long. She cut off a piece that was $1\frac{2}{3}$ yards long so that the remaining piece would be just the right size. Help Olivia determine how long her remaining piece is.

    a. Write each mixed number as an improper fraction.
    b. Determine which operation (add, subtract, multiply or divide) should be used.
    c. Follow the rules for that operation.
    d. Simplify your answer and convert it to a mixed number, if needed.

Step 4: Gregory's father told him he needed a rope that was $4\frac{1}{2}$ times the length of his arm. If Gregory's arm is $2\frac{1}{5}$ feet long, what length of rope does he need?

Step 5: In your own words, explain how to add, subtract, multiply and divide with mixed numbers.

EXAMPLE 1 Find the value of $2\frac{1}{4} + 1\frac{7}{8}$.

SOLUTION

Change each mixed number to an improper fraction.

$2\frac{1}{4} = \frac{9}{4}$ $1\frac{7}{8} = \frac{15}{8}$

Write equivalent fractions with the least common denominator, 8.

$\frac{9}{4} = \frac{18}{8}$ and $\frac{15}{8}$ (×2)

Add the numerators.

$\frac{18}{8} + \frac{15}{8} = \frac{33}{8}$

Write as a mixed number.

$\frac{33}{8} = 4\frac{1}{8}$

$2\frac{1}{4} + 1\frac{7}{8} = 4\frac{1}{8}$

Whole numbers can be turned into improper fractions by writing the whole number with 1 in the denominator. For example: $5 = \frac{5}{1}$.

EXAMPLE 2 Madison had a rope that was 8 feet long. She cut off a piece for a friend that was $5\frac{3}{4}$ feet long. How much rope does she have left?

SOLUTION

Write the problem.

$8 - 5\frac{3}{4}$

Change each mixed or whole number to an improper fraction.

$8 = \frac{8}{1}$ $5\frac{3}{4} = \frac{23}{4}$

Write equivalent fractions with the least common denominator, 4.

$\frac{8}{1} = \frac{32}{4}$ (×4)

Subtract the numerators and simplify.

$\frac{32}{4} - \frac{23}{4} = \frac{9}{4} = 2\frac{1}{4}$

Madison has $2\frac{1}{4}$ feet of rope left.

Lesson 1.5 ~ Operations with Mixed Numbers

EXAMPLE 3 Find the value of $\left(2\frac{1}{6}\right)\left(\frac{3}{10}\right)$.

SOLUTION

Write each mixed number as an improper fraction.

$$2\frac{1}{6} = \frac{13}{6} \qquad \frac{3}{10} = \frac{3}{10}$$

Multiply the numerators and denominators.

$$\frac{13}{6} \cdot \frac{3}{10} = \frac{13 \cdot 3}{6 \cdot 10} = \frac{39}{60}$$

Simplify the fraction.

$$\frac{39}{60} \xrightarrow{\div 3}_{\div 3} \frac{13}{20}$$

$$\left(2\frac{1}{6}\right)\left(\frac{3}{10}\right) = \frac{13}{20}$$

When finding the reciprocal of a whole number, write the whole number as a fraction over 1 and then "flip" the fraction.

$$3 = \frac{3}{1} \xrightarrow{\text{Reciprocal}} \frac{1}{3}$$

EXAMPLE 4 Iona runs six days each week. Each of her runs is equal in length. If her total mileage for the week is $12\frac{3}{4}$ miles, how long was each run?

SOLUTION

Write the problem.

$$12\frac{3}{4} \div 6$$

Write each whole and mixed number as an improper fraction.

$$12\frac{3}{4} = \frac{51}{4} \qquad 6 = \frac{6}{1}$$

Multiply by the reciprocal of the divisor.

$$\frac{51}{4} \div \frac{6}{1} = \frac{51}{4} \cdot \frac{1}{6} = \frac{51}{24}$$

Simplify the fraction.

$$\frac{51}{24} \xrightarrow{\div 3}_{\div 3} \frac{17}{8}$$

Change into a mixed number.

$$\frac{17}{8} = 2\frac{1}{8}$$

Iona runs $2\frac{1}{8}$ miles each day.

You can also cross reduce before multiplying.

$$\frac{\overset{17}{\cancel{51}}}{4} \cdot \frac{1}{\underset{2}{\cancel{6}}} = \frac{17}{8} = 2\frac{1}{8}$$

EXERCISES

1. In your own words, explain the process of multiplying two mixed numbers.

2. Yolanda decided that she did not need to change the two numbers in the problem below into improper fractions in order to add them. Do you agree or disagree? Why?
$$5 + 2\tfrac{1}{5}$$

Find each sum, difference, product or quotient. Write your answer in simplest form.

3. $3\tfrac{1}{2} + 1\tfrac{3}{4}$
4. $\left(2\tfrac{1}{2}\right)\left(4\tfrac{1}{2}\right)$
5. $4\tfrac{1}{3} - 1\tfrac{1}{4}$
6. $4\tfrac{3}{7} - 2\tfrac{6}{7}$
7. $3\tfrac{1}{3} - 1\tfrac{5}{6}$
8. $4\tfrac{3}{8} \div 1\tfrac{1}{4}$
9. $4\left(2\tfrac{3}{8}\right)$
10. $3\tfrac{3}{5} + \tfrac{7}{10}$
11. $7\tfrac{1}{3} \div 4$
12. $1\tfrac{3}{4} + 5\tfrac{1}{8}$
13. $\tfrac{14}{15} \div 2\tfrac{1}{3}$
14. $\left(3\tfrac{3}{4}\right)\left(1\tfrac{3}{5}\right)$
15. $6\tfrac{1}{2} - 2\tfrac{4}{5}$
16. $7\tfrac{1}{5} \div 1\tfrac{4}{5}$
17. $\left(5\tfrac{3}{4}\right)\left(3\tfrac{1}{2}\right)$

18. You have a piece of wire that is $6\tfrac{3}{4}$ inches long. You need to cut the wire into two pieces. One piece needs to be $4\tfrac{1}{4}$ inches long and the other piece needs to be $2\tfrac{2}{3}$ inches long.
    a. Write $4\tfrac{1}{4}$ and $2\tfrac{2}{3}$ as improper fractions.
    b. Find the sum of the improper fractions.
    c. Do you have enough wire? Explain.

19. Mia is making chocolate chip cookies. The recipe calls for $2\tfrac{2}{3}$ cups sugar. If Mia is tripling the recipe, how much sugar will she need?

20. During a weekend of rain in Savannah, the city recorded $1\tfrac{3}{4}$ inches the first day and $1\tfrac{1}{2}$ inches the second day. How many inches of rain did Savannah receive over the two-day weekend?

21. A telephone company has to install $2\tfrac{1}{4}$ miles of wire. If it takes a crew one day to install $\tfrac{3}{8}$ mile of wire, how long will it take the crew to install the entire $2\tfrac{1}{4}$ miles of wire? Explain how you know your answer is correct.

22. Jennifer long jumps $9\tfrac{1}{2}$ feet. Her friend, Sarah, long jumps $8\tfrac{7}{12}$ feet. How much further does Jennifer long jump than Sarah? Show all work necessary to justify your answer.

23. A hallway is $10\tfrac{1}{2}$ feet long and $2\tfrac{1}{3}$ feet wide. Half of the hallway is carpeted and the other half is hardwood. What is the area of the carpeted section? Use mathematics to justify your answer.

24. Zach tells his teacher that he can add and subtract mixed numbers without changing them into improper fractions.

a. Look at Zach's work below to find the value of $2\frac{1}{4} + 1\frac{3}{8}$. Explain Zach's process for adding the two mixed numbers.

$$2 + 1 = 3$$
$$\frac{1}{4} + \frac{3}{8} \rightarrow \frac{2}{8} + \frac{3}{8} = \frac{5}{8}$$
$$3 + \frac{5}{8} = \boxed{3\frac{5}{8}}$$

b. His teacher agrees that mixed numbers can be added or subtracted without changing them into improper fractions, but says that changing mixed numbers into improper fractions might be easier sometimes. The teacher gives an example of $4\frac{2}{3} + 1\frac{5}{6}$. Why do you think it might be easier to do this operation with improper fractions?

REVIEW

Find each sum, difference, product or quotient. Write your answer in simplest form.

25. $\frac{3}{4} + \frac{5}{12}$

26. $\frac{2}{5} - \frac{1}{6}$

27. $\frac{4}{11} \cdot \frac{1}{8}$

28. $\frac{5}{8} \div \frac{1}{4}$

29. $\frac{1}{4} - \frac{1}{4}$

30. $\frac{2}{3} \cdot \frac{7}{10}$

TIC-TAC-TOE ~ GRID GAME

You will create a grid game that other students can use to review fraction and decimal operations (+, −, ×, ÷). On a sheet of white paper, create an 8-inch by 8-inch grid that is divided into squares that are 2 inches on each side. Create a mixture of fraction and decimal sums, differences, products and quotients. Write an expression on an inside edge of a small square. Write the value of the expression on the adjacent square as seen below.

Continue creating fraction and decimal expressions that require the use of one operation. Put the corresponding answers on the adjacent square. Continue until all inside edges have either an expression or answer. If you would like, you may create distracters on the outside edges of the grid that do not have corresponding expressions or answers. Once your grid game is completed, cut the pieces apart and ask a friend to try to put the puzzle back together. Turn in a master copy of the grid game so your teacher could use it in the future.

ADDING AND SUBTRACTING DECIMALS

LESSON 1.6

 Add and subtract decimals.

Sam, Kim and Jackie got onto an elevator together. Kim weighs 45.36 kilograms. Jackie weighs 47.9 kilograms and Sam weighs 58.22 kilograms. The small elevator has a weight limit of 200 kilograms. One more person wants to get on the elevator but does not want to put them over the limit. How much can this person weigh without putting the elevator over the 200 kilogram limit?

When adding or subtracting decimals, you must line up the decimal points and then add or subtract the numbers as if they are whole numbers. When finished, you must bring down the decimal point.

Kim	45.36
Jackie	47.90
Sam	+ 58.22
	151.48 kilograms

Insert a zero at the end so all numbers have the same number of digits after the decimal point.

To find how much the last person can weigh, subtract the total weight of the three people from 200 kilograms.

```
 200.00
−151.48
  48.52 kilograms
```

If the decimal is not written, as in 200, it is at the far right of the number. Insert zeros after the decimal point to match the other numbers in the operation.

The last person who gets on the elevator can weigh a maximum of 48.52 kilograms.

ADDING AND SUBTRACTING DECIMALS

1. Line up the decimal points vertically.
2. Insert zeros so each number has the same amount of digits after the decimal point.
3. Add or subtract as with whole numbers.
4. Bring down the decimal point.

EXAMPLE 1   Find the value of 3.704 + 17.58.

SOLUTION   Write so that the decimals are lined up vertically.

Insert zeros so that there are the same number of digits after the decimal point.

```
  3.704
+17.580
 21.284
```

Move the decimal point down into the answer.

3.704 + 17.58 = 21.284

Lesson 1.6 ~ Adding and Subtracting Decimals   **25**

EXAMPLE 2 | Sophia is sick with the flu. There are 5.68 ounces of medicine in the bottle. Her mother plans to give her 0.88 ounces of the medicine today. How much will she have left?

SOLUTION

Write the problem. 5.68 − 0.88

Line up the decimal points and subtract.

```
  5.68
− 0.88
  4.80 ounces
```

It is not necessary to write a zero at the end of a decimal. 4.80 = 4.8

Her mother will have 4.8 ounces of medicine left at the end of the day.

The Commutative and Associative Properties of Addition allow you to rearrange and regroup numbers in addition problems.

PROPERTIES OF ADDITION

Commutative Property - The order in which numbers are added does not change the value of the expression.
$$5 + 3 = 3 + 5$$

Associative Property - The way in which numbers are grouped in addition expressions does not change the value of the expression.
$$2 + (4 + 7) = (2 + 4) + 7$$

EXAMPLE 3 | Walter went to the store and purchased the items listed below. What was the total cost for the four items?

Jar of Pickles	$2.30
Can of Beans	$1.60
Bottle of Juice	$1.15
Loaf of Bread	$1.70

SOLUTION

Using the Associative and Commutative Properties, regroup the prices so that mental math can be used.

$$2.30 + 1.70 = \$4.00$$
$$1.60 + 1.15 = \$2.75$$

Total = $4.00 + 2.75 = \$6.75$

Walter spent $6.75 at the store.

EXERCISES

Find each sum or difference.

1. 6.4 + 2.71

2. 9.45 − 1.8

3. 3.29 − 0.62

4. 14 + 2.39

5. 10.083 + 1.91

6. 4 − 1.93

7. 6.46 − 1.786

8. 106.045 + 3.06

9. 17.9 − 2.37

10. 0.49 + 10.1

11. 11.3 − 9

12. 14 − 1.71

13. John bought 2.64 pounds of broccoli and 3.95 pounds of carrots so his mom could make stir fry for dinner. What was the total weight of his purchase?

14. You decide to take your friend out to lunch. Your meal costs $6.34 and her meal costs $5.19.
  a. What was the total amount of the bill?
  b. If you give the cashier $15, how much change will you get back?

15. Raynesha walked 16.5 miles in the first week of January. In the second week, she walked 10.92 miles and in the third week she walked 12 miles. If she wants to reach a total of 60 miles in four weeks, how far does she need to walk in the last week? Show all work necessary to justify your answer.

16. Narineh has recorded her times in the 100 m dash for each of the first five track meets of the season. Use the table at the right to determine the difference between her fastest and slowest times.

17. Brendan purchased three small presents for his mother for her birthday and wants to determine the total amount he spent. One present cost $4.40. Another cost $5.98. The last present cost $1.60. Explain how to find the total amount Brendan spent on the gifts using mental math only.

Week	Time (Seconds)
1	14.13
2	13.9
3	14.38
4	13.22
5	14

REVIEW

Find each sum, difference, product or quotient. Write your answer in simplest form.

18. $2\frac{3}{4} + 2\frac{1}{3}$

19. $\left(5\frac{1}{2}\right)\left(\frac{3}{5}\right)$

20. $\frac{7}{12} - \frac{1}{4}$

21. $1\frac{3}{7} - \frac{6}{7}$

22. $\frac{2}{3} + 1\frac{5}{6}$

23. $3\frac{1}{8} \div 1\frac{1}{4}$

24. $6 \cdot \frac{4}{9}$

25. $5\frac{1}{4} - 2\frac{3}{8}$

26. $5\frac{1}{2} \div 10$

Lesson 1.6 ~ Adding and Subtracting Decimals

MULTIPLYING AND DIVIDING DECIMALS

LESSON 1.7

 Multiply and divide decimals.

All marathons are the same length. Patrick Makau from Kenya set a new world record for the marathon in 2011. He ran for about 2.06 hours at approximately 12.72 miles per hour. How long is the marathon?

In order to find the distance traveled by Makau, multiply his rate of 12.72 miles per hour by the amount of time he ran ($d = rt$).

$$d = (12.72)(2.06)$$

Multiply the two numbers as if they are whole numbers.

```
   12.72
  × 2.06
   7632
   0000
   2544
  262032
```

Count the number of digits after the decimal point in the two factors:
  12.72 → 2 decimal places
  2.06 → 2 decimal places

Move the decimal this many places in the product. Start on the right and move left.

26.2032

Round the answer to the nearest tenth.
Marathon ≈ 26.2 miles

MULTIPLYING WITH DECIMALS

1. Multiply as if the numbers are whole numbers.
2. Add the number of digits after the decimal point in each factor to find the number of decimal places in the product.
3. Insert zeros, if necessary, to hold place value. Delete zeros that are not necessary.

EXAMPLE 1 Find the value of 4.13(5.2).

SOLUTION Multiply as if the numbers are whole numbers and then insert the decimal point.

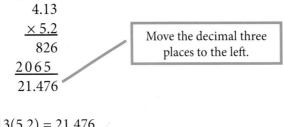

```
   4.13
  × 5.2
    826
   2065
  21.476
```

4.13(5.2) = 21.476

28 Lesson 1.7 ~ Multiplying and Dividing Decimals

When dividing decimals, you may first need to move the decimal point. If the divisor is not a whole number, the decimal point must be moved to the right until it is a whole number. The decimal point in the dividend must be moved the same number of places.

> **DIVIDING WITH DECIMALS**
> 1. If the divisor is not a whole number, change the divisor into a whole number by moving the decimal point to the right.
> 2. If the decimal point in the divisor was moved, move the decimal point in the dividend the same number of places to the right as it was moved in the divisor.
> 3. Divide the dividend by the divisor.
> 4. Move the decimal point into the quotient directly above the decimal point in the dividend.
> 5. If necessary, insert zeros to hold place values.

EXAMPLE 2 Find the value of 15.12 ÷ 3.6.

SOLUTION

Move decimals so the divisor is a whole number.

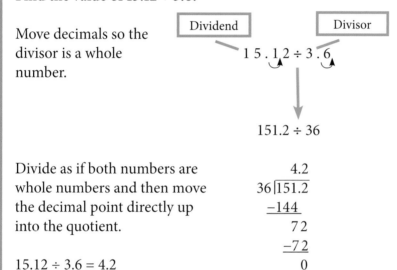

Divide as if both numbers are whole numbers and then move the decimal point directly up into the quotient.

```
       4.2
36 )151.2
    −144
      72
     −72
       0
```

15.12 ÷ 3.6 = 4.2

EXAMPLE 3 Isaiah developed 30 pictures from his digital camera. He was charged $2.70. What price did he pay per picture?

SOLUTION To find the price per picture, find the value of the total cost divided by the number of pictures developed:

2.70 ÷ 30

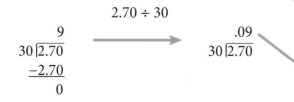

Move the decimal up into the quotient and insert zero to hold the place value.

Isaiah paid $0.09 per picture.

Lesson 1.7 ~ Multiplying and Dividing Decimals

EXAMPLE 4 Find the value of 5 ÷ 8.

SOLUTION

```
     .625
  8)5.000
   -48
     20
    -16
     40
    -40
      0
```

Insert zeros at the end of the dividend until there is a remainder of 0. The decimal point is placed in the quotient before the first zero was inserted.

5 ÷ 8 = 0.625

EXERCISES

Find each product.

1. 6.2(3)   **2.** 0.4(3.9)   **3.** 8(0.61)

4. 1.8(12)   **5.** 1.45(5.2)   **6.** 0.4(10.5)

7. 9(2.01)   **8.** 0.07(3.1)   **9.** 7.2(15.6)

10. Kendra's cell phone company charges her $0.29 per minute if she goes over her allotted minutes for the month. Last month she went over by 14 minutes. How much extra did she owe on her bill?

11. Kenyan bought 3.4 pounds of nails for his construction job. Each pound cost $1.59. How much did he pay for the nails? Round to the nearest cent.

12. The price tag under the 10.5 ounce bag of pretzels at the grocery store says the pretzels are $0.18 per ounce. How much do three bags of these pretzels cost? Show all work necessary to justify your answer.

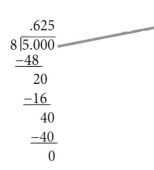

Find each quotient.

13. 10.64 ÷ 5.6   **14.** 8)29.6   **15.** 3.87 ÷ 0.9

16. 8.84 ÷ 6.8   **17.** 625.2 ÷ 12   **18.** 1.6)40

19. Shawna bought 2.8 pounds of bananas for $1.68. How much did the bananas cost per pound?

20. Orin spent $12.40 at the arcade. Skyler spent $18.60. How many times greater was Skyler's spending than Orin's spending?

21. Happy Sands Trucking delivers sand to families with sandboxes. The driver delivered 113.6 cubic feet of sand to 8 different houses. Each house received the same amount of sand. If the sand costs $4.10 per cubic foot, how much will each family pay for their sand? Use words and/or numbers to show how you determined your answer.

REVIEW

Find each sum or difference.

22. $2.48 + 3.9$

23. $0.87 - 0.043$

24. $90.472 - 5.81$

25. $6.04 + 12.2$

26. $7 - 4.32$

27. $9 + 124.7$

Tic-Tac-Toe ~ Memory Game

Write fifteen fraction and decimal expressions. Make sure you have a minimum of three expressions for each operation (+, −, ×, ÷). Cut heavy paper (such as card stock, construction paper, index cards or poster board) into 30 equal-sized pieces. Write each expression on one card and write the answer to the expression on another card in simplest form. Use these cards to play a memory game with a friend, classmate or family member. Record each pair of cards each participant wins on a sheet of paper. Turn in the cards and the game sheet to your teacher.

Tic-Tac-Toe ~ Numbers Galore

In many real-world situations, you often have to add more than two fractions or decimals to find the solution. There are two methods you can use to find the sums of expressions with more than two numbers:
(1) Find the sum of the first two numbers, then add the third number to the sum, then add the fourth number, and so on…
(2) Find the sum of all numbers at the same time by finding the least common denominator of all fractions or lining up the decimal points of all decimals.

Choose either method above to find the value of each expression. Write each answer in simplest form.

1. $\frac{1}{2} + \frac{3}{4} + \frac{5}{8}$

2. $\frac{4}{5} + \frac{1}{4} + \frac{3}{10}$

3. $\frac{1}{9} + \frac{1}{3} + \frac{1}{18}$

4. $5.49 + 0.04 + 18.2$

5. $102.03 + 52 + 8.9$

6. $0.005 + 0.17 + 3.458$

7. $\frac{3}{4} + \frac{3}{8} + \frac{1}{2} + \frac{5}{16}$

8. $3.2 + 0.08 + 10.9 + 2.07$

9. $1\frac{1}{2} + 2\frac{3}{4} + 3\frac{7}{8}$

10. Which method did you use to add the fraction expressions? Which method did you use to add the decimal expressions? Explain why you chose the method(s) you did.

Lesson 1.7 ~ Multiplying and Dividing Decimals

REVIEW

BLOCK 1

 Vocabulary

equivalent fractions   improper fraction   mixed number
fraction   least common multiple   reciprocal
greatest common factor least common denominator   simplest form

  Write fractions in simplest form.
Write mixed numbers as improper fractions.
Write improper fractions as mixed numbers.
Add and subtract fractions with like and unlike denominators.
Multiply and divide fractions.
Add, subtract, multiply or divide mixed numbers.
Add and subtract decimals.
Multiply and divide decimals.

Lesson 1.1 ~ Simplifying Fractions

Write two fractions that are equivalent to the given fraction.

1. $\frac{6}{9}$
2. $\frac{5}{20}$
3. $\frac{6}{10}$

Find the greatest common factor (GCF) of each pair of numbers.

4. 5 and 10
5. 8 and 12
6. 24 and 36

Write each fraction in simplest form.

7. $\frac{10}{20}$
8. $\frac{12}{30}$
9. $\frac{4}{16}$

10. $\frac{9}{15}$
11. $\frac{2}{22}$
12. $\frac{24}{54}$

13. Lyle said the length of his puppy's tail is $\frac{10}{12}$ foot long. Simplify this measurement.

14. Karianne bought $\frac{24}{32}$ pound of chicken. Simplify this measurement.

Lesson 1.2 ~ Mixed Numbers and Improper Fractions

Write the improper fraction and mixed number for each drawing.

15.

16.

Write each mixed number as an improper fraction.

17. $4\frac{1}{3}$

18. $2\frac{3}{5}$

19. $1\frac{8}{9}$

20. Anya's mother is making cookies for the holidays. Her recipe calls for $1\frac{3}{4}$ cups of flour. Write this amount as an improper fraction.

Write each improper fraction as a mixed number.

21. $\frac{9}{4}$

22. $\frac{22}{5}$

23. $\frac{25}{2}$

24. Nicki has grown $\frac{11}{4}$ inches in the last year. Hank has grown $2\frac{3}{4}$ inches. Who has grown more? Use mathematics to support your answer.

Lesson 1.3 ~ Adding and Subtracting Fractions

Find the least common multiple (LCM) of the numbers given.

25. 2 and 8

26. 8 and 12

27. 6 and 10

Find each sum or difference. Write your answer in simplest form.

28. $\frac{3}{10} + \frac{1}{10}$

29. $\frac{3}{4} - \frac{1}{8}$

30. $\frac{3}{5} + \frac{3}{4}$

31. $\frac{5}{6} + \frac{7}{10}$

32. $\frac{1}{2} - \frac{3}{10}$

33. $\frac{1}{3} + \frac{1}{2}$

34. Santiago vacuumed for $\frac{1}{3}$ hour and then swept floors for $\frac{1}{4}$ hour. For what fraction of an hour was he doing chores?

35. Tina brought home two kittens from the animal shelter. Slader weighed $\frac{7}{8}$ pound and Blaze weighed $\frac{7}{12}$ pound. How much heavier is Slader than Blaze? Explain how you know your answer is correct.

Block 1 ~ Review   33

Lesson 1.4 ~ Multiplying and Dividing Fractions

Find each product. Write the answer in simplest form.

36. $\frac{1}{3} \cdot \frac{1}{4}$

37. $\frac{3}{7} \cdot \frac{1}{6}$

38. $\frac{7}{10} \cdot \frac{5}{3}$

39. $\frac{1}{2} \cdot \frac{6}{11}$

40. $\frac{3}{4} \cdot \frac{2}{9}$

41. $\frac{5}{6} \cdot \frac{4}{5}$

42. Gary has a tarp that is $\frac{11}{12}$ yard long. He wants to cut off a piece that is $\frac{1}{3}$ of the size. How long will this piece of tarp be?

Find each quotient. Write the answer in simplest form.

43. $\frac{1}{5} \div \frac{1}{3}$

44. $\frac{3}{7} \div \frac{6}{7}$

45. $\frac{2}{9} \div \frac{1}{5}$

46. $\frac{7}{10} \div \frac{3}{4}$

47. $\frac{2}{5} \div \frac{2}{11}$

48. $\frac{9}{10} \div \frac{3}{8}$

49. Ivan has a board that is $\frac{5}{8}$ yard long. He plans to cut the board into smaller boards that are each $\frac{5}{32}$ yard long. How many boards will he be able to cut? Explain how you know your answer is correct.

Lesson 1.5 ~ Operations with Mixed Numbers

Find each sum, difference, product or quotient. Write the answer in simplest form.

50. $2\frac{1}{2} + 3\frac{1}{4}$

51. $(3\frac{1}{3})(1\frac{1}{5})$

52. $4\frac{1}{5} - 3\frac{1}{4}$

53. $1\frac{3}{7} - \frac{5}{7}$

54. $5 - 2\frac{7}{10}$

55. $9\frac{3}{4} \div 3\frac{1}{4}$

56. $7(1\frac{3}{5})$

57. $1\frac{4}{5} + 6\frac{9}{10}$

58. $5\frac{1}{4} \div 3$

59. Paul is making chili. The recipe calls for $1\frac{3}{4}$ cups of canned tomatoes. If Paul is tripling the recipe, how many cups of canned tomatoes will he need?

60. During a weekend in Junction City, the city recorded $3\frac{2}{3}$ inches of snow on Saturday and $1\frac{1}{2}$ inches of snow on Sunday. How many inches of snow will they need to get on Monday if they want to break a three-day snow record of $10\frac{3}{4}$ total inches? Show all work necessary to justify your answer.

Lesson 1.6 ~ Adding and Subtracting Decimals

Find each sum or difference.

61. 5.3 + 2.58   **62.** 23.25 − 0.8   **63.** 8.7 − 0.53

64. 14 + 7.39   **65.** 11.3 + 5.81   **66.** 7 − 2.43

67. Ethan participates in a triathlon. It includes swimming, biking and running. The swimming portion is 0.8 miles. The biking portion of the triathlon is 12.25 miles.
   a. How far has Ethan gone after the swimming and biking portions?
   b. The entire triathlon is 15.6 miles. How long is the running portion?

68. A local news station hosted a three-day fundraiser for flood victims after a winter storm in Missouri. On the first day they received $1,405.62. On the second day they collected $912.44. They want to raise a total of $4,000 for the whole fundraiser. How much do they need to receive on the third day? Use words and/or numbers to show how you determined your answer.

Lesson 1.7 ~ Multiplying and Dividing Decimals

Find each product or quotient.

69. 7(5.4)   **70.** 1.2 ⟌ 13.2   **71.** 4.96 ÷ 0.8

72. 3.4(621)   **73.** 10.5 ÷ 0.84   **74.** 0.04(1.9)

75. The price tag below the 8.5-ounce box of crackers at the grocery store says the crackers are $0.28 per ounce. How much does the box of crackers cost?

76. Jed bought 2.24 pounds of pears and 1.36 pounds of apples for $5.22. If the pears and apples are the same price per pound, what is the price per pound for the fruit? Show all work necessary to justify your answer.

TIC-TAC-TOE ~ DECIMALS OR FRACTIONS?

Write a play with at least four characters. One character should be a fraction. Another character should be a decimal. In the play, the fraction and decimal argue about whom the easiest to work with is when performing the basic operations of adding, subtracting, multiplying and dividing. The characters should provide multiple reasons and examples for their way of thinking.

CAREER FOCUS

JEFF
CHEF

I am a professional chef. I do a wide variety of tasks in my job. I plan for events, prepare meals and clean up after an event is over. I also prepare budgets and decide how many staff I will need to make sure that an event goes as planned.

Math skills are important in all aspects of my daily duties. I use ratios and proportions to make sure I have the correct amount of food to prepare each dish. I need to know how many portions need to be prepared for a specific meal. I do not want to make too much and end up wasting food. I often use formulas to figure out how much to buy and prepare.

Another important part of being a chef is knowing different types of measurement. Some foods are weighed by ounces while others might be measured in gallons or pounds. I use percentages to calculate how much money to pay for items as well as how much to charge. My calculations have to be accurate to make sure people get just what they want and our company makes a good profit.

I went to a two-year training program in order to become a chef. There are many culinary programs available that do a good job preparing students for this field. I began working as a chef after getting my training. I have gained experience and different skills as I have worked over the years.

Salaries for chefs vary depending upon what type of restaurant or company you work for and how much experience you have. A chef can expect to make around $30,000 per year when they start their career. With experience and skill, chefs can earn much more than that.

I like my job because I enjoy creating meals that people enjoy. It is very rewarding to go through the whole process of buying raw products, preparing them and presenting them to customers. My job changes almost every day and I enjoy working with the other people on my culinary team.

CORE FOCUS ON RATIONAL NUMBERS & EQUATIONS

BLOCK 2 ~ INTEGERS

Lesson 2.1	Understanding Integers	39
Lesson 2.2	Adding Integers	44
	Explore! Integer Chips	
Lesson 2.3	Subtracting Integers	49
Lesson 2.4	Multiplying Integers	52
	Explore! Number Jumping	
Lesson 2.5	Dividing Integers	56
Lesson 2.6	Powers and Exponents	60
	Explore! Positive or Negative?	
Lesson 2.7	Order of Operations	64
	Explore! Fact Puzzle	
Review	Block 2 ~ Integers	68

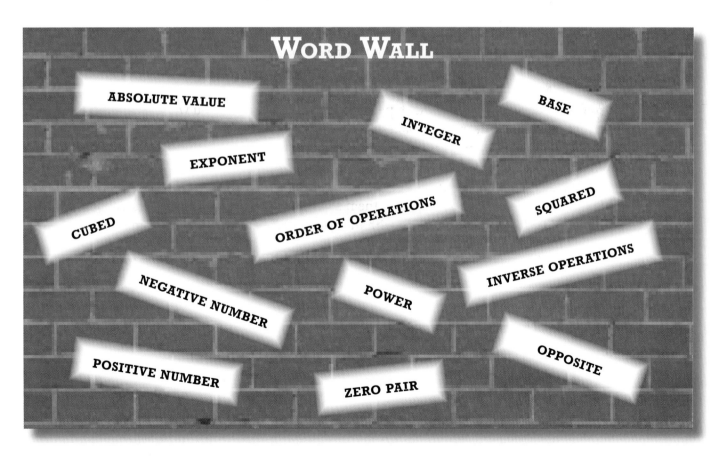

BLOCK 2 ~ INTEGERS
TIC-TAC-TOE

EQUALS 30 Write an expression using a given set of integers so that it equals 30. *See page 67 for details.*	**INTEGER BINGO** Design an integer BINGO game which requires players to find the value of integer operations. *See page 59 for details.*	**CARD GAME** Create a card game where players use black and red cards to represent integers. *See page 59 for details.*
VIDEO GAME Write a proposal for a video game where players earn integer point values based on situations. *See page 48 for details.*	**INTEGER MODELS** Create a brochure that explains how to perform integer operations using integer chips or number lines. ●●● + ●●●●● *See page 58 for details.*	**INTEGER EXPONENTS** Find the value of powers with integer exponents. Summarize your findings. *See page 63 for details.*
CHILDREN'S STORY Write a children's story about integers. The main character will encounter integers in real-world situations. *See page 71 for details.*	**HIGHS AND LOWS** Research the record high and low temperatures for 10 different states. *See page 51 for details.*	**HISTORY OF INTEGERS** Research and write a paper about the history of integers. *See page 55 for details.*

UNDERSTANDING INTEGERS

LESSON 2.1

Graph integers and their opposites on a number line.
Compare and order integers.

Numbers used to represent values less than zero are called **negative numbers**. **Positive numbers** are any numbers greater than 0.

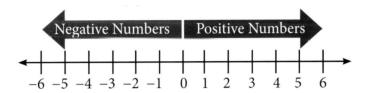

Two numbers are **opposites** if they are the same distance from 0 on a number line, but on opposite sides of 0. The opposite of 4, which is a positive number, is −4. The number −4 is read "negative four". **Integers** are the set of all positive whole numbers, their opposites and zero.

EXAMPLE 1 Find and graph the opposite of each integer.
a. 2   b. −6

SOLUTIONS

a.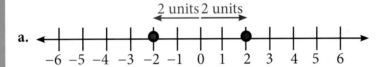

The opposite of 2 is −2 because each integer is 2 units from 0, but in opposite directions.

b.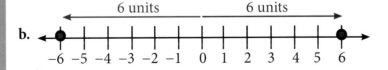

The opposite of −6 is +6 (which is written as 6) because each integer is 6 units from 0, but in opposite directions.

The **absolute value** of a number is the distance that number is from 0 on a number line. The absolute value of −8 is written $|-8|$. The absolute value of a number is always positive. For example, $|-8| = 8$.

Lesson 2.1 ~ Understanding Integers 39

EXAMPLE 2 Find each absolute value.
a. $|-4|$ b. $|3|$

SOLUTIONS

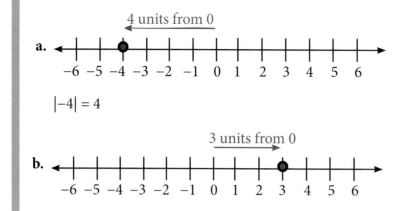

$|-4| = 4$

$|3| = 3$

Integers are used in everyday situations. Positive integers are associated with gains and increases. Negative integers are used when describing losses or decreases. Negative integers are also used to describe something that is below sea level while positive integers describe something above sea level.

EXAMPLE 3 Write an integer to represent each situation.
a. Death Valley, California is 282 feet below sea level.
b. The stock market gained 5 points yesterday.
c. Patrice withdrew $15 from her savings account.

SOLUTIONS

a. Death Valley, California is 282 feet <u>below</u> sea level.
Answer: −282

b. The stock market <u>gained</u> 5 points yesterday.
Answer: 5

c. Patrice <u>withdrew</u> $15 from her savings account.
Answer: −15

On a number line, the numbers get larger as you move from left to right. On the number line below, the integer represented by the letter A is the smallest integer marked on the number line. Point E represents the largest integer marked on the number line.

$A < B < C < D < E$

SMALLEST ⟶ LARGEST

40   Lesson 2.1 ~ Understanding Integers

EXAMPLE 4   Complete each statement using < or >.

a. −4 ◯ 0   b. −3 ◯ −5   c. −6 ◯ 7

SOLUTIONS

a.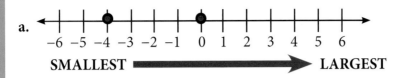

Since −4 is to the left of 0 on the number line, −4 < 0.

Remember:
< means less than
> means more than

b. Since −3 is to the right of −5 on the number line, −3 > −5.
c. Because −6 is to the left of 7 on the number line, −6 < 7.

EXAMPLE 5   The five states with the lowest temperatures on record are shown in the table below. List and graph the temperatures from least to greatest.

Record Low Temperatures

Location	Lowest Temperature (°F)
Prospect Creek, Alaska	−80
Peter's Sink, Utah	−69
Maybell, Colorado	−61
Roger's Pass, Montana	−70
Riverside, Wyoming	−66

SOLUTION   When the temperatures are graphed on the number line, the lowest temperature will be furthest to the left. The highest temperature will be the furthest to the right.

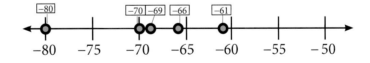

Listed from least to greatest: −80, −70, −69, −66, −61

EXERCISES

Find the opposite of each number.

1. 8

2. −1

3. −7

4. 1.6

5. −109

6. $\frac{3}{4}$

Lesson 2.1 ~ Understanding Integers   41

7. Draw a number line that goes from −5 to 5. Graph the number 2 and its opposite.

8. Draw a number line that goes from −5 to 5. Graph the number −4 and its opposite.

9. Draw a number line that goes from −10 to 10. Graph the following integers on the number line: 2, −5, 9, −1, and 4.

10. Which of the following are integers? 6, −2, $3\frac{1}{3}$, −10, −2.7

11. Name one number that is a negative number but is not a negative integer. Explain your answer.

Find each absolute value.

12. $|9|$ **13.** $|-1|$ **14.** $|-14|$

15. $\left|2\frac{1}{4}\right|$ **16.** $\left|\frac{8}{9}\right|$ **17.** $|-6.8|$

18. What two numbers both have an absolute value of 7? Explain how you know your answer is correct.

Write an integer to represent each situation.

19. The basement floor is 12 feet below the ground.

20. Three people joined a group for lunch.

21. J.T. owes his friend $8.

22. The record low temperature in North America is 81°F below zero in the Yukon Territory.

23. Leticia moved forward 2 spaces on a board game.

24. At 11,239 feet, Mount Hood is the tallest peak in Oregon.

25. Create a situation that would be represented by a positive integer. Explain why a positive integer makes sense for your situation.

26. Create a situation that would be represented by a negative integer. Explain why a negative integer makes sense for your situation.

27. Integers have been graphed on a number line. Explain how you can tell which number is the smallest and which number is the largest by looking at the number line.

Copy and complete each statement using < or >.

28. 9 ⬤ 11 **29.** 5 ⬤ 0 **30.** 0 ⬤ −2

31. −8 ⬤ −3 **32.** −7 ⬤ 4 **33.** −22 ⬤ −25

Order each set of integers from least to greatest.

34. 3, 0, 4, −1

35. 5, 2, −2, 0, −1

36. −4, −10, 4, −12, −6

37. −3, −9, −12, −1, −5

38. −73, −21, −40, −29

39. 17, 21, −17, −21, 0

40. A six-story building has stairs that are each 7 inches tall. Dave, who works on the top floor of the building, says that the change in height for each stair should be represented by −7. Carol, who works on the first floor, thinks that the change in height should be represented by +7. Who do you think is correct? Support your answer with your reasoning.

41. Name five integers between −4 and 4. Order them from least to greatest.

Copy and complete each statement using <, > or =.

42. |6| ● |7|

43. |−4| ● |−5|

44. |8| ● |−8|

45. Jules put a list of numbers in order from least to greatest. His list is below.

$$-1, -3, -7, -10, 0, 1, 4, 9$$

  a. Did he order them correctly? If not, explain his mistake.
  b. If necessary, correctly order the numbers from least to greatest.

46. Maria, James and Paul played a card game where the lowest score wins. Maria scored −7 points and James scored −1 point. Paul won the game. Describe Paul's possible scores.

REVIEW

Find the value. Write your answer in simplest form.

47. $\frac{1}{3} + \frac{1}{6}$

48. $\frac{2}{3} \cdot \frac{6}{11}$

49. $\frac{5}{8} - \frac{1}{12}$

50. $\frac{3}{14} \div \frac{3}{7}$

51. $\frac{3}{10} + \frac{9}{20}$

52. $\left(1\frac{2}{3}\right)\left(\frac{1}{2}\right)$

ADDING INTEGERS

LESSON 2.2

 Add two or more integers to find the sum.

EXPLORE!   **INTEGER CHIPS**

Integer chips are helpful for modeling integer operations. Each blue chip will represent the integer +1. Each red chip will represent the integer −1. When a positive integer chip is combined with a negative integer chip, the result is zero. This pair of integer chips is called a **zero pair**.

$$+1 \; + \; -1 = 0$$

Step 1: Model 4 + (−2) with integer chips.

Step 2: Group as many zero pairs as possible.

Step 3: Because zero pairs are worth zero, remove all zero pairs. What is the result?

Step 4: Write an expression for the following model. Create the same model using your own integer chips.

Step 5: Group as many zero pairs as possible and remove them. What is the result of your expression?

Step 6: Model −5 + (−3) with integer chips.

Step 7: Are there any zero pairs? If so, remove them. What is the result of −5 + (−3)?

Step 8: Write an integer addition expression. Model it with integer chips to find the sum.

Step 9: Use integer chips to help you determine what type of answer (positive, negative or zero) you think you will get for each situation. Explain your reasoning.
   a. the sum of two positive numbers
   b. the sum of two negative numbers
   c. the sum of a number and its opposite
   d. the sum of a negative and positive integer

44 *Lesson 2.2 ~ Adding Integers*

Number lines are another way to model integer addition. Look at these integer sums on a number line:

You can find the value of 4 + (−2) using a number line.
- Start at 0
- Move 4 to the right for +4
- Move 2 to the left for −2

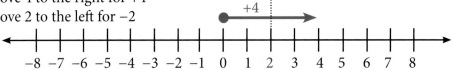

The value of 4 + (−2) = 2.

You can find the value of −5 + (−3) using a number line.
- Start at 0
- Move 5 to the left for −5
- Move 3 to the left for −3

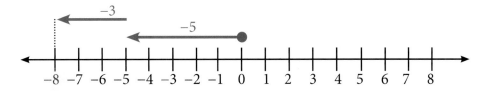

The value of −5 + −3 = −8.

EXAMPLE 1   Find the sum using integer chips or a number line.
a. −1 + (−4)   b. 6 + (−3)

SOLUTIONS   a. −1 + (−4)
Using integer chips: ● + ●●●● = ●●●●● = −5

Using a number line:

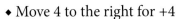

b. 6 + (−3)
Using integer chips:

Zero pairs are removed.

Using a number line:

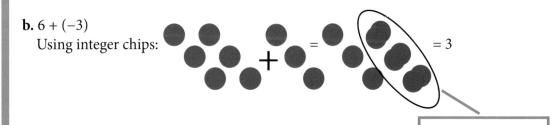

Lesson 2.2 ~ Adding Integers **45**

The following rules may help you add two integers without using integer chips or a number line.

> **ADDING INTEGERS**
>
> 1. Determine the sign of the answer:
>   - If both integers are positive, the sum is positive.
>   - If both integers are negative, the sum is negative.
>   - If one integer is negative and the other is positive, the sum is the sign of the number with the greater absolute value.
> 2. Find the sum:
>   - If the integers have the same sign, add the absolute value of the numbers and use the sign determined above.
>   - If the integers have different signs, subtract the lesser absolute value from the greater absolute value and use the sign determined above.

EXAMPLE 2 JaNell descended 16 feet and then rose 21 feet in elevation on a hike. Write an integer addition expression to represent this situation. Determine her overall change in elevation.

SOLUTION JaNell descended 16 feet and then rose 21 feet.
Integer expression: −16 + 21
Determine the sign.
  One integer is negative and the other is positive so use the sign of the number that has the larger absolute value. Since |21| > |−16|, the sign of the answer is positive.
Find the sum.
  Subtract the lesser absolute value from the larger absolute value.
  21 − 16 = 5

JaNell's overall change in elevation is 5 feet.

EXAMPLE 3 Find the value of −5 + (−2) + 1.

SOLUTION

| Find the sum of the first two numbers. | −5 + −2 = −7 |
| Add the third number to the sum. | −7 + 1 = −6 |

−5 + (−2) + 1 = −6

EXERCISES

Use a number line to find each sum.

1. −5 + (−1)   **2.** 6 + (−2)   **3.** −3 + 4

4. −9 + 7   **5.** −2 + (−3)   **6.** 8 + (−2)

Find each sum. Use integer chips, a number line or the integer addition rules.

7. −2 + (−7)   **8.** −3 + 9   **9.** −4 + 3

10. 14 + (−10)   **11.** −11 + (−1)   **12.** −5 + 5

13. 7 + (−8)   **14.** −15 + (−6)   **15.** 10 + 12

16. 100 + (−30)   **17.** −5 + 18   **18.** −25 + (−15)

19. Jason withdrew $14 from his account on Monday. He deposited $8 in his account on Tuesday. What integer represents the total change in his account over the last two days?

20. Maria is in a two-day golf tournament. She scored −3 on the first day. On the second day, her score is −5. What is her overall score for the entire tournament? Use two different methods to show your answer is correct.

21. Ishmael's stock went up $17 on Thursday and then down $13 on Friday. What was the total change in the value of the stock?

Find each sum.

22. −3 + (−4) + (−1)   **23.** 12 + (−5) + 3   **24.** 10 + (−7) + 6

25. 20 + 15 + (−12)   **26.** −8 + (−4) + 5   **27.** −1 + (−2) + (−3) + (−4)

28. Yesterday Kirk borrowed $13 from his friend. Today he borrowed $8 more. He plans to pay back his friend $15 tomorrow. What will Kirk's balance be? Show all work necessary to justify your answer.

29. Find three different integer addition expressions that equal −4. Use mathematics to justify your answers.

30. Santiago ran up 72 stairs. He realized he forgot something and went down 32 stairs. He finished by going up 65 more stairs.
 a. Write an integer addition expression to show Santiago's movement up and down the stairs.
 b. What was Santiago's ending location on the stairs?

Lesson 2.2 ~ Adding Integers

31. Esther opened a checking account by depositing $60. Her next four transactions are shown in the table. Copy and complete the "Integer" column of the table by writing an integer to represent each transaction. Find her balance at the end of this transaction period.

Transaction	Integer
Deposited $60	+60
Withdrew $22	
Withdrew $6	
Deposited $35	
Withdrew $20	
Balance	

32. Determine the sign of each answer without adding the values. Explain your reasoning.
   a. −3429 + 4502
   b. 10770 + 2980
   c. −60.86 + −902.5
   d. 0.067 + (−0.1)

REVIEW

Order the integers from least to greatest.

33. −5, 4, −1, 2

34. −1, −6, −10, −2, 0

35. 5, −3, −1, −2, −7

Find each absolute value.

36. |8|

37. |−12|

38. $\left|-1\frac{2}{3}\right|$

39. Carlina bought 1.52 pounds of raisins and 1.08 pounds of prunes for $6.50. The raisins and prunes cost the same price per pound. What is the price per pound for the dried fruit? Show all work necessary to justify your answer.

Tic-Tac-Toe ~ Video Game

Write a proposal for a kid-friendly video game. Players can earn integer point values based on specific situations in the game. Design a flyer giving the details of your game, including the following:
- the theme
- the characters
- all the different ways participants can earn positive integers
- all the different ways participants can earn negative integers

At the end of your flyer include an example level of your game. Explain what is happening to the character, how many points they are earning for each event and a score total after each event.

SUBTRACTING INTEGERS

LESSON 2.3

  Subtract two integers to find the difference.

The record high temperature in Florida is 109° F. The lowest temperature on record in Florida was −2° F. What is the difference between these temperatures?

$$109 - (-2) = ?$$

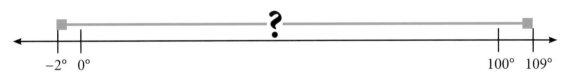

When subtracting integers, one method is to make the expression into an addition expression by adding the opposite. The difference in high and low temperatures in Florida can be found by adding the opposite.

Change to addition Change to its opposite
$$109 - (-2) = ?$$
$$109 + (+2) = 111° F$$

The difference between the highest and lowest temperatures in Florida is 111° F.

SUBTRACTING INTEGERS

1. Subtract the integers by adding the opposite.
2. Follow the rules for adding integers to determine the sum.

EXAMPLE 1 Use a number line to find the value of −2 − 3.

SOLUTION −2 − 3 is −2 − (+3)
Add the opposite and use the number line to find the value.   −2 + (−3) = −5

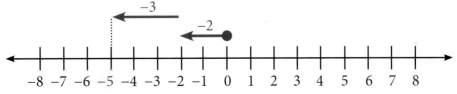

Lesson 2.3 ~ Subtracting Integers 49

EXAMPLE 2 | Use integer chips to find the value of −5 − (−4).

SOLUTION | Subtract the integers by adding the opposite: −5 + 4

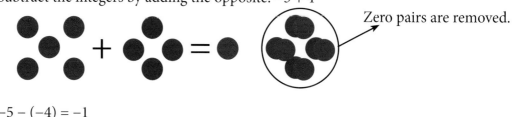

Zero pairs are removed.

−5 − (−4) = −1

EXAMPLE 3 | The melting point of mercury is approximately −39° C. The melting point of chlorine is approximately −101° C. How much higher is mercury's melting point than chlorine's melting point?

SOLUTION | Subtract the melting point of chlorine from the melting point of mercury.
−39 − (−101)
Add the opposite.   −39 + 101
Find the sign of the larger absolute value and then subtract.
101 − 39 = 62° C
Mercury's melting point is 62° C higher than chlorine's melting point.

EXERCISES

1. Explain how to change an integer subtraction expression into an integer addition expression. Give an example.

Use a number line to find each difference.

2. −4 − 2   **3.** 7 − (−1)   **4.** 5 − 4

5. 1 − 6   **6.** −3 − (−3)   **7.** 8 − (−2)

8. The current balance of Amber's bank account was $15. She withdrew $21. Find the new balance in her account.

Find each difference. Use integer chips, a number line or the integer subtraction and addition rules.

9. 8 − 2   **10.** 5 − (−3)   **11.** −9 − 4

12. −7 − 4   **13.** −9 − (−10)   **14.** 18 − (−2)

15. 40 − 15   **16.** 17 − 19   **17.** 6 − (−12)

18. −13 − 5   **19.** −3 − 10   **20.** −8 − (−8)

Lesson 2.3 ~ Subtracting Integers

21. At 20,320 feet above sea level, Mount McKinley in Alaska is the highest point in the United States. The lowest point in the United States is in Death Valley, California. Death Valley is 282 feet below sea level. Write a subtraction expression to determine the difference in elevation between the highest and lowest points in the United States. Find the difference.

22. Write a subtraction expression that involves two negatives and has a positive answer. Explain how you know your answer is correct.

Determine if each statement is always true, sometimes true or never true. Give a counterexample unless it is always true.

23. A positive number minus a positive number is a positive.

24. A negative number minus a negative number is a negative.

25. A negative number minus a positive number is a positive.

26. A positive number minus a negative number is a positive.

REVIEW

Find the value.

27. 4.1 + 5.08

28. 3(5.6)

29. 12.3 − 8.9

30. 6.4 ÷ 4

31. 7.28 + 3 + 6.8

32. 2.3(7.1)

Tic-Tac-Toe ~ Highs and Lows

Write a paper about the record high and low temperatures across the United States. Include the following in your paper:

Step 1: Predict which state(s) you believe might have the highest record temperature and which state(s) might have the lowest record temperature. Explain why you chose those states.

Step 2: Research the record high and low temperatures in at least 10 different states. Create a table and record the state, the highest and lowest temperatures recorded in that state and the date each occurred.

Step 3: Find the range (largest minus smallest) between the highest and lowest temperature in each of the states in **Step 2**. Did your findings surprise you? Why or why not?

MULTIPLYING INTEGERS

LESSON 2.4

 Multiply two integers to find the product.

Multiplication can be described as repeated addition. When you learned to multiply positive whole numbers, you may have learned that 4(2) is the same as saying "four groups of two".

EXPLORE!   **NUMBER JUMPING**

Step 1: Draw four number lines, each from −8 to 8.

Step 2: Use the first number line to model the solution to 4(2). Do this by starting at 0 and moving "four groups of two" to the right. Why do you think you moved to the right? What is the solution to 4(2)?

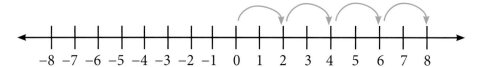

Step 3: Use the second number line to model the solution to 4(−2). Do this by starting at 0 and moving "four groups of negative two". Which direction will you be moving? What is the solution?

Step 4: Illustrate −4(2) on the third number line. Do this by starting at 0 and moving "the opposite of four groups of two". Which direction did you move to illustrate the "opposite of four groups of two"? What is the solution?

Step 5: On the last number line, show the solution to −4(−2). Start at 0 and move "the opposite of four groups of negative two." What is your solution? Explain to a classmate how you got your solution.

Step 6: Use the number lines to help you determine which type of answer (positive or negative) you think you will get for each situation. Explain your reasoning.
       a. The product of two positive integers.
       b. The product of a positive and negative integer.
       c. The product of a negative and positive integer.
       d. The product of two negative integers.

USE REPEATED ADDITION ON NUMBER LINES TO MULTIPLY INTEGERS.

EXAMPLE 1   Find each product.
a. 5(−4)   b. −2(−3)

SOLUTIONS

a. This product can be read "five groups of negative four". The integers have different signs so the product is negative. The solution is −20.

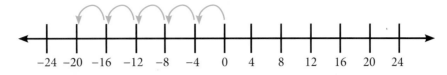

b. This product can be read "the opposite of two groups of negative three". Move two groups of negative three and then find the opposite. The integers have the same signs so the product is positive. The solution is 6.

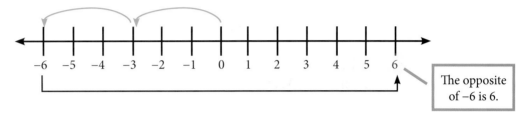

The opposite of −6 is 6.

MULTIPLYING INTEGERS

1. Determine the sign of the answer.

   ◆ The product of two integers with the same sign is positive.

   ◆ The product of two integers with different signs is negative.

2. Find the product of the absolute value of the numbers and use the sign determined above.

EXAMPLE 2   A submarine descends below the water's surface at a rate of 50 meters per minute. Find the integer that represents the submarine's position relative to the surface of the water after 6 minutes.

SOLUTION   Write an expression to represent the situation.   −50(6)

Negative because the submarine is descending.

The two integers have different signs so the product is negative.   −50(6) = −300 meters

The submarine's position after 6 minutes is −300 meters.

Lesson 2.4 ~ Multiplying Integers 53

Multiplying more than two integers together does not require special rules. It can be done by multiplying two integers first and then continuing to multiply the product by the other integers.

EXAMPLE 3   Find the value of 2(−5)(−4).

SOLUTION   Multiply the first two integers. $2(-5) = -10$
Multiply the product by the last integer. $-10(-4) = 40$

2(−5)(−4) = 40

EXERCISES

1. Can the same rules be used to determine the sign of an integer sum and an integer product? Give at least one example to prove your answer.

2. Explain how you can look at an integer multiplication expression and determine if the product is negative or positive.

Model each product using a number line.

3. 3(2)   **4.** 6(−1)   **5.** −3(−4)

Find each product. Use a number line or the integer multiplication rules.

6. 5(−8)   **7.** −1(1)   **8.** −4(−9)

9. −7(−2)   **10.** 6(−10)   **11.** −11(5)

12. 12(3)   **13.** −9(−2)   **14.** 6(−15)

15. −3(4)   **16.** 6(−1)   **17.** −3(−4)

18. 13(−2)   **19.** −4(−5)   **20.** 4(−5)

21. Allison withdrew $7 from her savings account every day. If she continued to do this for a total of 5 days, find the integer that represents the change in the value of her savings account.

22. José is making iced tea. He has ice cubes which he uses to cool the iced tea. Each ice cube lowers the temperature of the liquid by 2° F. He adds 5 ice cubes. What integer represents the change in temperature of the iced tea when he adds the ice cubes? Use a model to show that your answer is correct.

23. Curtis enjoys scuba diving with his family during the summer. When he dives into the water, he descends 4 feet per second. What integer represents his position relative to the water's surface after 9 seconds?

24. The price of a stock drops $3 each day for 7 days. The stock was worth $50 before the price began dropping. What is the price of the stock after the drop in price? Use words and/or numbers to show how you determined your answer.

Determine if each statement is always true, sometimes true or never true. Give a counterexample unless it is always true.

25. A positive number times a positive number is positive.

26. A negative number times a negative number is negative.

27. A negative number times a positive number is negative.

Find each product.

28. −3(5)(2) **29.** 6(−4)(−1) **30.** −2(−2)(−2)

31. −10(4)(3) **32.** −2(3)(5)(−10) **33.** −1(−1)(−3)(−6)

34. During a game, Demetrius had seven turns where he went back 2 units on each turn and three turns where he went forward 5 units on each turn. What integer represents the total number of units he moved? Show all work necessary to justify your answer.

35. Maggie borrowed $4 from one friend and $8 from another friend. She did this five days in a row. What integer represents the amount of money she owes? Show all work necessary to justify your answer.

REVIEW

Find each value. Use integer chips, a number line or the integer subtraction and addition rules.

36. −2 + 7 **37.** 8 − (−2) **38.** −9 + 4

39. −7 − 4 **40.** −9 + (−10) **41.** 1 − (−2)

42. 30 + (−12) **43.** 27 − 32 **44.** −6 + (−1)

45. Jonathan and Belinda were playing a board game. Belinda moved her game piece forward 7 spots from the starting point on her first turn. She moved backwards 5 spots on her next turn. On her third turn, she moved forward one spot. What integer represents her current position from the starting point on the board? Use mathematics to justify your answer.

Tic-Tac-Toe ~ History of Integers

Research the history of integers. When were negative numbers first used? Who invented integers? What were integers first used for? Write a 1–2 page paper explaining the history of integers. Include as much information as you can in an organized manner. Include a bibliography to cite your sources.

DIVIDING INTEGERS

LESSON 2.5

 Divide two integers to find the quotient.

Division and multiplication are inverse operations. Inverse operations are operations that undo each other. Look at the following multiplication and division fact families.

Multiplication Equations	Related Division Equations
$6(4) = 24$	$24 \div 6 = 4$ or $24 \div 4 = 6$
$-6(4) = -24$	$-24 \div (-6) = 4$ or $-24 \div 4 = -6$
$6(-4) = -24$	$-24 \div 6 = -4$ or $-24 \div (-4) = 6$
$-6(-4) = 24$	$24 \div (-6) = -4$ or $24 \div (-4) = -6$

The related multiplication and division expressions above show that when two integers have the same sign their quotients are positive. When two integers have different signs, their quotient is negative. These integer division rules follow the same rules that you followed when multiplying integers.

DIVIDING INTEGERS

1. Determine the sign of the answer.
    ♦ The quotient of two integers with the same sign is positive.
    ♦ The quotient of two integers with different signs is negative.
2. Find the quotient of the absolute values of the numbers and use the sign determined above.

EXAMPLE 1   Find each quotient.
     a. $40 \div (-5)$ **b.** $\dfrac{-20}{-2}$

SOLUTIONS   **a.** $40 \div (-5)$
The two integers have different signs so $+ \div - = -$
the quotient is negative.
Find the quotient of the absolute values $40 \div 5 = 8$
of the numbers.
Use the correct sign on the quotient. $40 \div (-5) = -8$
Check with the related multiplication fact. $-5(-8) = 40$

b. $\dfrac{-20}{-2}$ *Remember the fraction bar means division.*

The two integers have the same sign so $- \div - = +$
the quotient is positive.
Find the quotient of the absolute values $20 \div 2 = 10$
of the numbers.
Use the correct sign on the quotient. $\dfrac{-20}{-2} = 10$

EXAMPLE 2

Rita's family went hiking last weekend. They walked to the top of a waterfall and then turned around to come back down. It took them 15 minutes to descend 600 feet in elevation at a steady pace. What integer represents their change in elevation each minute?

SOLUTION

Since the family is descending 600 feet, the integer will be written as a negative number.

$$-600 \div 15 = -40$$

The family's elevation change is represented by −40 feet per minute.

EXERCISES

1. What are the similarities or differences in the integer multiplication and division rules?

2. What is a related multiplication expression for $12 \div (-2) = -6$?

Find each quotient.

3. $-8 \div 2$

4. $\frac{-30}{-6}$

5. $18 \div 9$

6. $28 \div (-4)$

7. $-80 \div (-10)$

8. $\frac{-36}{-9}$

9. $\frac{-8}{-1}$

10. $-35 \div 7$

11. $-56 \div (-8)$

12. $-6 \div 6$

13. $\frac{24}{4}$

14. $-36 \div (-6)$

15. $\frac{100}{-25}$

16. $-70 \div (-10)$

17. $18 \div (-3)$

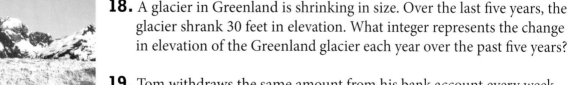

18. A glacier in Greenland is shrinking in size. Over the last five years, the glacier shrank 30 feet in elevation. What integer represents the change in elevation of the Greenland glacier each year over the past five years?

19. Tom withdraws the same amount from his bank account every week. He has withdrawn $60 over the past four weeks. What integer represents the weekly change in his bank account?

20. While playing a game with her brother, Lola lost points on four consecutive turns: 18, 12, 4 and 6. What integer represents her average loss on these four turns? Show all work to justify your answer.

Determine if each solution is positive or negative without doing any calculations.

21. 274(31)

22. −405 ÷ (−90)

23. 98(107)

24. −357 + 89

25. 602 + 481

26. −925 − 471

27. Blake told his teacher that he thinks the same rules that work for adding integers work for multiplying and dividing. The addition rules that he wants to use for multiplication and division are listed below. Is he correct? If not, describe which rules change for multiplication and division.
Rule 1: Positive and Positive = Positive
Rule 2: Negative and Negative = Negative
Rule 3: Positive and Negative = Either (depends on the sign of the larger number)

28. Write an expression using all four operations (+, −, ×, ÷) and at least one negative number. The value of the expression must be negative. Show all work to prove that your expression fits all the constraints.

REVIEW

Change each improper fraction into a mixed number in simplest form.

29. $\frac{21}{5}$

30. $\frac{17}{3}$

31. $\frac{12}{4}$

32. $\frac{32}{6}$

33. $\frac{42}{4}$

34. $\frac{27}{7}$

Change each mixed number into an improper fraction.

35. $3\frac{1}{2}$

36. $6\frac{2}{5}$

37. $5\frac{3}{4}$

38. $1\frac{5}{8}$

39. $6\frac{1}{3}$

40. $2\frac{5}{6}$

Tic-Tac-Toe ~ Integer Models

In this block you learned how to perform some integer operations using models (integer chips or number lines). Create a brochure explaining how to use integer chips for one operation (+, −, ×, ÷) and how to use the number line for a different operation. Include a second operation for each type of model. On one panel of the brochure, include information about why models may be helpful for some students.

Tic-Tac-Toe ~ Integer BINGO

Create a worksheet that has 6 of each type of integer operations (sums, differences, products and quotients). Allow room for others to put their answers on the worksheet. On a separate sheet type the answers to the integer expressions. No two answers can be the same. Create a blank BINGO card with a free space in the center of the card. The cards should have 5 rows and 5 columns.

To play the game, participants will find the value of the expressions and place their answers anywhere on the BINGO card. The person in charge of the activity will call out the answers to expressions in no particular order. Participants cross off the boxes with the corresponding answers. The game can be played for a normal BINGO and then continued on for a blackout. All participants that get a blackout will have found the value of all 24 expressions correctly.

Lead your BINGO activity with at least three other people. Once you have completed the activity, collect their BINGO cards and turn them in to your teacher along with your worksheet, answer key and blank BINGO card.

Tic-Tac-Toe ~ Card Game

Use a regular deck of playing cards for this activity. Take out all the face cards (Jacks, Queens and Kings). Each black numbered card represents one of the positive integers (for example, ♣4 = +4). Each red numbered card represents one of the negative integers (for example, ♥7 = −7).

Create a card game that can be played with two people. The card game must make the players use integer operations to win "points". Be creative with your rules.

Some ideas to think about:
- Does each player get half the cards? Five cards? Ten cards?
- Are players trying to find pairs that add, subtract, multiply or divide to a certain amount?
- Does one player get points if the sum is negative and the other player get points if the sum is positive?
- Is there one card that is "wild"?

Once you have designed your game, ask two different pairs of people to try it out. Ask each player to write a short review of your game once they have played it. Read the reviews and write a one-page paper summarizing the feedback. Also include in your paper any changes you would make in the rules before the game was played again. Turn in your paper along with an updated set of rules for your game.

POWERS AND EXPONENTS

LESSON 2.6

 Write and compute expressions with powers.

When an expression is the product of a repeated factor, it can be written using a **power**. A power consists of two parts, the **base** and the **exponent**. The base of the power is the repeated factor. The exponent shows the number of times the factor is repeated.

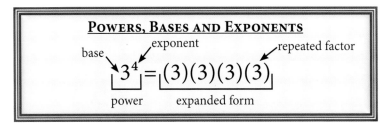

It is important to know how to read powers correctly.

Power	Reading the Expression	Expanded Form	Value
5^2	"five to the second power" or "five **squared**"	(5)(5)	25
$(-4)^3$	"negative four to the third power" or "negative four **cubed**"	(-4)(-4)(-4)	-64
3^4	"three to the fourth power"	3 · 3 · 3 · 3	81

EXPLORE! POSITIVE OR NEGATIVE?

Step 1: Copy the table below.

Base	Base Squared	Base Cubed	Base to the Fourth Power	Base to the Fifth Power
-2				
-1				
1				
2				

Step 2: Fill in the table by writing each power in expanded form. Find the value of the power. The table has been started for you below.

Base	Base Squared	Base Cubed	Base to the Fourth Power	Base to the Fifth Power
-2	$(-2)^2$ (-2)(-2) = 4	$(-2)^3$ (-2)(-2)(-2) = -8		

EXPLORE! (CONTINUED)

Step 3: Some of the answers are positive and some answers should be negative. Do you notice any patterns that would help you determine the sign of a power just by looking at the base and the exponent?

Step 4: Use your pattern to determine the SIGN of each power. You do not need to find the value.
   a. 8^5 b. $(-9)^4$ c. $(-6)^{11}$
   d. $(-4)^8$ e. 12^2 f. $(-1)^{17}$

Step 5: The expressions -3^2 and $(-3)^2$ are not equal. Explain why the two expressions are not equal and find the value of each expression.

EXAMPLE 1

Write the numerical expression as a power.
   a. $(5)(5)(5)(5)$ b. $(-11)(-11)$ c. $7 \cdot 7 \cdot 7 \cdot 7 \cdot 7 \cdot 7$

SOLUTIONS
   a. 5^4 b. $(-11)^2$ c. 7^6

POWER RULES FOR POSITIVE AND NEGATIVE BASES

- If the base of a power is positive, the value of the power will be positive.
- If the base of a power is negative, the value of the power will be:
  - positive if the exponent is an even number
  - negative if the exponent is an odd number.

EXAMPLE 2

Write each power in expanded form and find the value.
   a. 8^2 b. $(-1)^5$ c. -4^4

SOLUTIONS

   a. Expanded Form: $8 \cdot 8$
   Value: 64

   b. Expanded Form: $(-1)(-1)(-1)(-1)(-1)$
   The exponent is odd so the answer will be negative.
   Value: -1

   c. Expanded Form: $-(4)(4)(4)(4)$
   The expression -4^4 should be read "the opposite of four to the fourth power".
   $4^4 = 256$ so the opposite of 256 is -256.
   Value: -256

EXERCISES

1. Using the power $(-2)^3$, what number represents the:
   a. base?
   b. exponent?
   c. value of the power?

2. Explain in words the difference between $(-5)^2$ and -5^2.

Write the numerical expression as a power.

3. $9 \cdot 9 \cdot 9 \cdot 9$

4. $(-2)(-2)(-2)$

5. $(10)(10)$

6. $(-3)(-3)(-3)(-3)(-3)(-3)$

7. $(8)(8)(8)(8)(8)$

8. $12 \cdot 12 \cdot 12$

Determine if the value of each power will be positive or negative.

9. 15^3

10. $(-22)^6$

11. -7^2

12. $(-3)^9$

13. $(-38)^4$

14. $(-11)^7$

Write each power in expanded form and find the value.

15. 4^2

16. $(-1)^5$

17. $\left(\frac{1}{2}\right)^2$

18. 10^2

19. 3^4

20. $(-2)^3$

21. $(-7)^2$

22. $\left(\frac{2}{3}\right)^3$

23. 4^5

Copy and complete each statement using <, > or =.

24. $(-2)^2$ ⬤ -2^2

25. 3^2 ⬤ $(-3)^2$

26. 5^2 ⬤ 2^5

27. The volume of a cube can be calculated using the formula $v = lwh$.
   a. Write the volume of the cube shown in expanded form.
   b. Write the volume of the cube as a power.
   c. Find the value of the volume of the cube.

28. Maggie planned a 40th birthday party for her mother. She called six of her mother's friends and asked them to each call six more friends. Those people each called six more people.
   a. Write the number of contacted people as a sum of powers.
   b. How many people were contacted in all?

REVIEW

Find the value of each integer expression.

29. $50 + (-30)$ **30.** $-8(2)$ **31.** $24 \div (-3)$

32. $-7 - 2$ **33.** $-4(-5)$ **34.** $-1 + (-5)$

35. $2 - 12$ **36.** $-50 \div (5)$ **37.** $6(-7)$

38. $3 + 43$ **39.** $6 - (-1)$ **40.** $-3(-9)$

Tic-Tac-Toe ~ Integer Exponents

1. Copy the table below on the top half of a piece of paper. Use a calculator to fill in the value of each power. Convert each value to a fraction in simplest form.

   For example: $5^2 = 25 = \frac{25}{1}$ or $5^{-2} = 0.04 = \frac{4}{100} = \frac{1}{25}$

Power	Value of Power (decimal)	Value of Power (fraction)
2^1		
2^2		
2^3		
2^4		
2^{-4}		
2^{-3}		
2^{-2}		
2^{-1}		

2. On the same sheet of paper, summarize your findings about negative exponents. Use your findings to predict the values of each power as a fraction. Check your answers using a calculator and fill in the "Value of Power (decimal)" column.

Power	Value of Power (decimal)	Value of Power (fraction)
5^1		
5^{-1}		
3^2		
3^{-2}		
4^2		
4^{-2}		
9^2		
9^{-2}		

ORDER OF OPERATIONS

LESSON 2.7

 Find the value of expressions using the order of operations.

The 12 members of Weston Junior High's Karate Team want to enter a karate tournament. There is a team entry fee of $100 and an additional fee of $8 per person. Their coach writes an equation that will allow them to figure out the total cost to enter the tournament.

$$\text{Cost} = 100 + 12(8)$$

One team member believes it will cost the team $896 to enter the tournament. Another team member disagrees. He believes it will cost $196. Look at their calculations below. Who do you agree with?

Team Member #1
100 + 12(8)
= 112(8)
= $896

Team Member #2
100 + 12(8)
= 100 + 96
= $196

Mathematicians have established an **order of operations**. The order of operations are rules which should be followed when evaluating an expression with more than one operation. Using the correct order of operations would help the karate team calculate the total cost to enter the tournament. In this situation, Team Member #2 was correct. It will cost $196 for the team to enter the karate tournament.

The order of operations for expressions includes grouping symbols, powers, multiplication, division, addition and subtraction. All operations inside grouping symbols must be completed before any of the other steps are completed. Two types of grouping symbols you will learn how to work with are parentheses and the fraction bar.

> **ORDER OF OPERATIONS**
> 1. Find the value of expressions inside grouping symbols such as parentheses and fraction bars.
> 2. Find the value of all powers.
> 3. Multiply and divide from left to right.
> 4. Add and subtract from left to right.

EXAMPLE 1 Find the value of each expression.

a. $4(-5 + 3) - 7$

b. $27 - (1 + 3)^2 \div 2$

SOLUTIONS

a. Add the integers inside the parentheses.　　　　$= 4(-2) - 7$
Multiply 4 and −2.　　　　$= -8 - 7$
Add the opposite.　　　　$= -15$

b. Add the integers inside the parentheses.　　　　$= 27 - (4)^2 \div 2$
Find the value of 4^2.　　　　$= 27 - 16 \div 2$
Divide 16 and 2.　　　　$= 27 - 8$
Subtract 8 from 27.　　　　$= 19$

EXAMPLE 2 Find the value of each expression.

a. $\dfrac{-5(-7) + 14}{2 + 5}$

b. $\dfrac{(6 - 9)^2}{2 - 3} + 5$

SOLUTIONS

a. Multiply in the numerator. 　　　　$\dfrac{-5(-7) + 14}{2 + 5} = \dfrac{35 + 14}{2 + 5}$

Add the values in the numerator.　　　　$= \dfrac{49}{2 + 5}$

Add the values in the denominator.　　　　$= \dfrac{49}{7}$

Divide the numerator by the denominator.　　　　$= 7$

b. Add the opposite in the parentheses.　　　　$\dfrac{(6 - 9)^2}{2 - 3} + 5 = \dfrac{(-3)^2}{2 - 3} + 5$

Find the value of $(-3)^2$.　　　　$= \dfrac{9}{2 - 3} + 5$

Add the opposite in the denominator.　　　　$= \dfrac{9}{-1} + 5$

Divide the numerator by the denominator.　　　　$= -9 + 5$

Add the integers.　　　　$= -4$

Lesson 2.7 ~ Order of Operations

EXPLORE!

FACT PUZZLE
Source: www.infoplease.com

Use the order of operations to find the answers to these facts.

Step 1: The first motion picture theatre was opened in Los Angeles, California in what year?

$$100(-3 + 18) + 8 \cdot 50 + 2$$

Step 2: Cleveland, Ohio was home to the first traffic light. What year was it installed?

$$\frac{60 \cdot 30}{4 - 3} - 2(-57)$$

Step 3: Philadelphia, Pennsylvania was the first city to have a daily newspaper. When was it first printed?

$$2000 - (10 + 5)^2 + 3^2$$

Step 4: The first subway was built in Boston, Massachusetts. What year was the subway built?

$$\frac{12}{4 - 8} + 40(-2 + 22) + 10^3 + 10^2$$

Step 5: The first elevator was built in New York, New York. What year was the elevator built?

$$1800 - 2(2)(-13)$$

EXERCISES

Find the value of each expression.

1. $\dfrac{8 + 12}{6 - 1}$

2. $7(1 + 3) - 6$

3. $(-2 + 3)^2 - 8$

4. $\dfrac{(3 + 5)^2}{2} + 2(5 - 11)$

5. $-20(2 + 2) - 10$

6. $\dfrac{-6(4)}{(-3 + 1)^3}$

7. $\dfrac{22 - 5(2 + 1)}{-3 + (-4)}$

8. $5(7 - 1)^2 - 10 \cdot 2$

9. $5 + \dfrac{3 + (-7)}{2}$

10. $\dfrac{12 + 4}{6 - 4} + \dfrac{6^2}{9}$

11. $(4 - 2)^3 - 2 \cdot 4$

12. $-7(3 \cdot 4 - 7)$

13. $18 + 2(-5 - 10)$

14. $\dfrac{5 \cdot 9 - 10}{5 - (-2)}$

15. $(10 - 3)^2 + (-1)^2$

Lesson 2.7 ~ Order of Operations

16. The Booster Club sold tickets to Saturday's basketball tournament. Admission with Student ID was $2 per person. Regular admission was $5. The booster club sold 120 student tickets and 50 regular tickets. How much money did the Booster Club collect? Show all work necessary to justify your answer.

17. Three friends order a pizza for $14, a two-liter bottle of soda for $3 and a tub of cookie dough for $4. They want to evenly split the total cost. Determine how much each person owes. Use words and/or numbers to show how you determined your answer.

18. Explain why it is necessary to have an order of operations in mathematics.

19. Create an expression with at least five numbers and two different operations that has a value of 12. Use mathematics to justify your answer.

Insert one set of parentheses in each numerical expression so that it equals the stated amount. Use mathematics to justify your answer.

20. $5 + 3 + 9 \div 3 = 9$

21. $15 + 1 \cdot 4 - 2 \cdot 3 = 58$

REVIEW

Write each power in expanded form and find the value.

22. 6^2

23. $(-3)^3$

24. 12^2

25. $\left(\frac{1}{2}\right)^2$

26. -5^2

27. $(-1)^4$

Find the value of each integer expression.

28. $6 + (-2)$

29. $-7(-9)$

30. $\frac{50}{-25}$

31. $-8 - (-7)$

32. $-4 + (-5)$

33. $-1(-11)$

Tic-Tac-Toe ~ Equals 30

Use all the integers in each set below to find an expression that equals 30. You can use addition, subtraction, multiplication and/or division. You may also use parentheses to group numbers, if needed. Check the answers using the order of operations.

1. −5, 3, −2

2. 7, 40, −3

3. 1, 4, −5, 5

4. 6, −4, 10, 2

5. 8, −1, −3, 6

6. 3, 6, 7, −9

7. −5, −15, 9, −1

8. −3, 5, 2, −1

REVIEW

BLOCK 2

Vocabulary

absolute value	integer	positive number
base	inverse operations	power
cubed	negative number	squared
exponent	opposite	zero pair
	order of operations	

Graph integers and their opposites on a number line.
Compare and order integers.
Add two or more integers to find the sum.
Subtract two integers to find the difference.
Multiply two integers to find the product.
Divide two integers to find the quotient.
Write and compute expressions with powers.
Find the value of expressions using the order of operations.

Lesson 2.1 ~ Understanding Integers

Find the opposite of each number.

1. 6

2. –9

3. –11

4. Draw a number line that goes from –5 to 5. Graph the number –3 and its opposite.

Find each absolute value.

5. |7|

6. |–3|

7. |–46|

Write an integer to represent each situation.

8. The buried treasure is 18 feet below the ground.

9. Cameron owes his mother $5.

10. Quinn's stock increased by $9.

Copy and complete each statement using < or >.

11. 9 ⬤ 11

12. 5 ⬤ 0

13. 0 ⬤ –2

Order the integers from least to greatest.

14. 5, 0, –4, 1   **15.** 7, –3, –2, 0, 2   **16.** –1, –9, –6, –12, –3

17. Use the table below.

Technogram Stock Account Activity

Week	Activity	Integer
1	Increased by 5 points	
2	Dropped 3 points	
3	Gained 10 points	
4	Fell 7 points	
5	Decreased by 6 points	

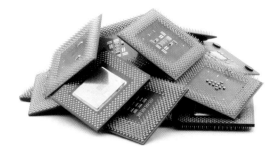

  a. Copy and complete the "Integer" column of the table by using an integer to describe each week's activity.
  b. List the integers that represent the stock activity from least to greatest.

Lesson 2.2 ~ Adding Integers

Find each sum. Use integer chips, a number line or the integer addition rules.

18. –5 + (–3)   **19.** –6 + 2   **20.** 10 + (–4)

21. 11 + 10   **22.** –20 + (–3)   **23.** –4 + 4

24. 7 + (–6)   **25.** –31 + (–2)   **26.** –1 + 9

27. Rebecca withdrew $22 from her account on Thursday. She deposited $10 in her account on Friday. What integer represents the total change in her account over those two days?

28. Mitchell was in a two-day golf tournament. His score was +4 on the first day. On the second day, he scored –6. What was his overall score for the entire tournament? Show two ways to find the answer.

Lesson 2.3 ~ Subtracting Integers

Find each difference. Use integer chips, a number line or the integer subtraction and addition rules.

29. 11 – 3   **30.** 9 – (–4)   **31.** –5 – 2

32. –10 – 1   **33.** –8 – (–10)   **34.** 21 – (–4)

35. 7 – 15   **36.** 29 – 19   **37.** 70 – (–31)

38. Victor dove off a high dive that was 30 feet above the water. When he dove he was able to touch the bottom of the pool 10 feet below the water's surface. What is the difference between the height of the board and the lowest point of his dive? Explain how you know your answer is correct.

Lesson 2.4 ~ Multiplying Integers

Find each product. Use a number line or the integer multiplication rules.

39. 5(−6) **40.** −1(4) **41.** −9(−9)

42. −10(−2) **43.** 6(−8) **44.** −12(3)

45. 9(5) **46.** −5(−3) **47.** 7(−11)

48. Nigel withdrew $15 from his savings account every week. Find the integer that represents the change in value of his savings account after 6 weeks.

49. Leilah made soup at a homeless shelter. The soup was too hot to serve. Leilah used ice cubes to cool the soup. Each ice cube lowered the temperature of the liquid by 3° F. She added 4 ice cubes to the soup. What integer represents the change in temperature to the soup if she did this twice? Show all work necessary to justify your answer.

Lesson 2.5 ~ Dividing Integers

Find each quotient.

50. −16 ÷ 8 **51.** $\frac{-35}{-5}$ **52.** 28 ÷ 7

53. 40 ÷ (−2) **54.** −99 ÷ (−11) **55.** $\frac{15}{-3}$

56. $\frac{-8}{-8}$ **57.** −42 ÷ 6 **58.** −24 ÷ (−4)

59. A small airplane is 2,000 feet above the ground. The pilot begins to descend in order to land the plane. It takes the pilot 4 minutes to reach the ground. Assuming the plane is descending at a constant speed, what integer represents the plane's change in elevation per minute?

60. Amadeus lost 40 points over the last five rounds of a card game he is playing with his friends. He lost the same number of points during each round. Find the integer that represents the amount of points he lost per round.

Lesson 2.6 ~ Powers and Exponents

Write the numerical expression as a power.

61. $4 \cdot 4 \cdot 4 \cdot 4 \cdot 4$

62. $(-7)(-7)(-7)$

63. $(1)(1)(1)(1)$

Write each power in expanded form and find its value.

64. 8^2

65. $(-6)^2$

66. $\left(\frac{1}{3}\right)^2$

67. 12^2

68. $(-3)^3$

69. -1^4

Copy and complete each statement using <, > or =.

70. $(-3)^2 \bigcirc -3^2$

71. $4^2 \bigcirc 2^4$

72. $6^2 \bigcirc 3^3$

Lesson 2.7 ~ Order of Operations

Find the value of each expression.

73. $5(4 + 2) - 3$

74. $\dfrac{-3 + 31}{6 - (-1)}$

75. $(9 + 2)^2 - 9 \cdot 3$

76. $\dfrac{(4 + 6)^2}{4} + 3(4)$

77. $-10(-7 + 11) + 5$

78. $\dfrac{-8(2)}{(-3 + 1)^3}$

79. The circus in town last weekend was sold out. They sold 200 youth tickets for $4 each and 140 adult tickets for $10 each.
   a. Write an expression to represent the total amount of money the circus brought in last weekend.
   b. How much money did the circus make in ticket sales?

80. Five siblings bought take-out for their family's dinner. They split three meals. The meals cost $14, $12 and $9. They divided the total cost equally among them. How much did dinner cost for each sibling? Show all work necessary to justify your answer.

Tic-Tac-Toe ~ Children's Story

Integers are used in many real-world experiences. Create a children's book that incorporates the concept of integers. The character(s) in your book should encounter integers in a variety of real-world situations. The plot should include the character(s) solving integer sums, differences, products or quotients. Your book should have a cover, illustrations and a story line that is appropriate for children.

CAREER FOCUS

Kim
Pharmacist

I am a pharmacist. I work in a pharmacy located inside a hospital. My responsibilities include giving medications to patients and monitoring medication profiles to be sure patients receive appropriate care. I work with doctors, nurses and other health professionals to ensure patients get the best possible therapy in our center. Much of a pharmacist's job is providing information and recommendations about different medications, especially making sure people get the right dosage of medicine. It also includes making sure patients are aware of any allergies they might have to different medications. Pharmacists can also work in retail stores, clinics or mail order pharmacies.

Pharmacists use math daily. I calculate doses of medications based on information such as the height and weight of a patient. Some patients require special calculations to make sure their medicines are dosed correctly. For example, children get different amounts of medications than adults do. Some medicines, called antibiotics, require constant monitoring to make sure the dose is high enough to treat an infection, but not high enough to cause harm to the patient. I use math to interpret data and adjust dosages in those situations.

A pharmacist must have a doctor of pharmacy degree from a college or school of pharmacy. An applicant to such a program must have completed at least two years of college which include courses in math and science. Most pharmacy programs also require applicants to take an admissions test. I had to pass a national pharmacy exam as well as a law exam in order to get my license after graduating. Continuing education is required because new medical information and medications are constantly being developed.

A new pharmacy graduate can expect to earn $85,000 to $95,000 per year. On average, pharmacists earn between $96,000 and $117,000 per year.

I love my job. I have the opportunity to help people in my community when they are very sick. I live in a small town and know many of the patients who are treated in our hospital. I am proud that I have used my special skills to make sure people get the correct dose of medications in the safest way possible.

CORE FOCUS ON RATIONAL NUMBERS & EQUATIONS

BLOCK 3 ~ RATIONAL NUMBER OPERATIONS

Lesson 3.1	Estimating Sums and Differences	75
	Explore! Trip to the Store	
Lesson 3.2	Adding Rational Numbers	80
Lesson 3.3	Subtracting Rational Numbers	84
	Explore! What's the Difference?	
Lesson 3.4	Estimating Products and Quotients	88
	Explore! In Your Head	
Lesson 3.5	Multiplying Rational Numbers	92
Lesson 3.6	Dividing Rational Numbers	97
Review	Block 3 ~ Rational Number Operations	101

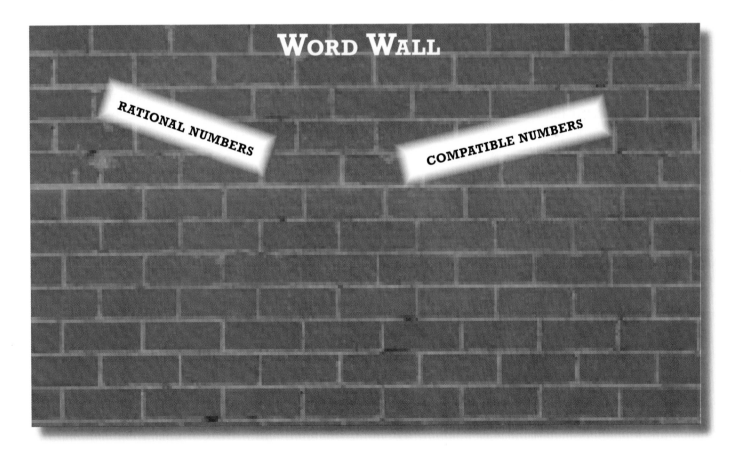

BLOCK 3 ~ RATIONAL NUMBER OPERATIONS

TIC-TAC-TOE

Scientific Notation

Learn to write very large and very small numbers in scientific notation.

See page 96 for details.

Irrational Numbers

Research and write a report about irrational numbers.

See page 79 for details.

Stocks

Follow three stocks for a week. Show daily changes in value and total change over a week-long period of time.

See page 87 for details.

Tutorial

Create a tutorial for adding and subtracting rational numbers. Write a quiz for students to take after completing the tutorial.

See page 104 for details.

Order of Operations

Evaluate rational number expressions using the order of operations.

See page 100 for details.

Estimation Worksheet

Create a 20-question worksheet to help other students practice estimation techniques.

See page 91 for details.

Classifications of Numbers

Create a poster that illustrates the different classifications of numbers.

See page 95 for details.

Estimation Poetry

Write two poems about situations where it may or may not be helpful to estimate rational numbers.

See page 79 for details.

Chemical Elements

Examine four different chemical elements. Find information about each element.

See page 91 for details.

ESTIMATING SUMS AND DIFFERENCES

LESSON 3.1

 Estimate sums and differences of rational numbers.

A number that can be written as a fraction of two integers is called a **rational number**. Integers, fractions and some decimals are types of rational numbers. The integer addition and subtraction rules you learned in **Block 2** apply to all rational numbers. In this lesson, you will use these rules while estimating sums and differences of rational numbers.

RATIONAL NUMBERS

Integers
Fractions
Some Decimals
Mixed Numbers

EXPLORE!   TRIP TO THE STORE

A family of four went to the grocery store. The family split up the list of items to purchase. Help the family estimate as they find the items they need.

Step 1: The father picks up three items in the frozen food section. His items cost $4.79, $5.07 and $9.49. He wants to determine the approximate cost of his three items. Help him by rounding each number to the nearest dollar and then adding the prices together.

Step 2: Janessa, the daughter, was asked to buy three different kinds of apples. She weighed each type. She got $\frac{2}{5}$ pound of Granny Smith apples, $\frac{3}{16}$ pound of Golden Delicious apples and $\frac{11}{12}$ pound of Gala apples. She wants to determine the approximate total weight of the apples.
   a. Determine if each fraction is closer to 0, $\frac{1}{2}$ or 1. Round each fraction to the number it is closest to.
   b. Add the estimated weights together.

Step 3: Mario, the son, bought four different-sized candy bars. The candy bars cost $0.89, $0.59, $0.19 and $0.69. Mario wants to determine the approximate cost for the candy bars.
   a. Determine if each decimal is closer to 0, 0.5 or 1. Round the cost of each candy bar to one of these numbers.
   b. Add the estimated amounts together.

Step 4: The mother went to the bulk food aisle. She purchased $2\frac{3}{4}$ pounds of oatmeal, $4\frac{1}{8}$ pounds of corn meal and $1\frac{5}{16}$ pounds of peanuts. After putting the items in her basket, she wants to determine the total weight of the food in her basket. Round each mixed number to the nearest whole number. Add these estimated amounts together.

Lesson 3.1 ~ Estimating Sums and Differences 75

EXPLORE! (CONTINUED)

Step 5: Why do you think people would estimate in the grocery store?

Step 6: Describe a situation where you estimated with fractions or decimals.

ESTIMATING SUMS AND DIFFERENCES

DECIMALS
1. With numbers less than one unit from zero, round to 0, ± 0.5 or ± 1.
2. With numbers that are more than one unit from zero, round to the nearest integer.

FRACTIONS
1. With fractions less than one unit from zero, round to 0, $\pm \frac{1}{2}$ or ± 1.
2. With mixed numbers that are more than one unit from zero, round to the nearest integer.

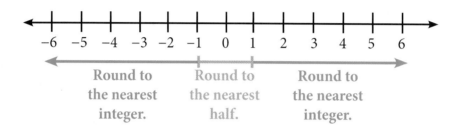

EXAMPLE 1

Use estimation to simplify each expression.
a. −4.23 + 2.8
b. 0.94 − (−0.43)

SOLUTIONS

a. Both numbers are more than one unit from zero. Each number should be rounded to the nearest integer and then added.

$-4.23 \approx -4$
$2.8 \approx 3$
$-4 + 3 = -1$

≈ is a symbol for "approximately"

$-4.23 + 2.8 \approx -1$

b. Both numbers are less than one unit from zero so the numbers should be rounded to the nearest 0.5.

Add the opposite.
$0.94 \approx 1$
$-0.43 \approx -0.5$

$1 - (-0.5) = 1 + 0.5 = 1.5$

$0.94 - (-0.43) \approx 1.5$

EXAMPLE 2

Estimate the value of each expression.
a. $-\frac{1}{7} + \frac{4}{9}$
b. $-1\frac{1}{3} - 2\frac{7}{8}$

SOLUTIONS

a. Since each fraction is less than 1 unit from zero, round each to the nearest $\frac{1}{2}$.

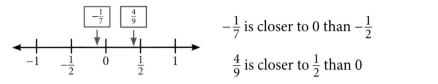

$-\frac{1}{7}$ is closer to 0 than $-\frac{1}{2}$ $-\frac{1}{7} \approx 0$

$\frac{4}{9}$ is closer to $\frac{1}{2}$ than 0 $\frac{4}{9} \approx \frac{1}{2}$

$0 + \frac{1}{2} = \frac{1}{2}$

$-\frac{1}{7} + \frac{4}{9} \approx \frac{1}{2}$

b. Mixed numbers should be rounded to the nearest integer.

$-1\frac{1}{3} \approx -1$

$2\frac{7}{8} \approx 3$

Add the opposite. $-1 - 3 = -1 + (-3) = -4$

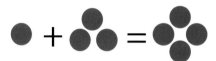

$-1\frac{1}{3} - 2\frac{7}{8} \approx -4$

When rounding fractions less than one unit from zero, it is helpful to consider this information:
- If the numerator is very small compared to the denominator, it should be estimated as 0.
- If the numerator is about half of the denominator, it should be estimated as $\frac{1}{2}$ or $-\frac{1}{2}$.
- If the numerator is nearly as big as the denominator, it should be estimated as -1 or 1.

EXERCISES

1. Give two real-world situations where estimation may be used.

2. Write a two-number addition expression where:
   a. both numbers would be rounded to the nearest integer.
   b. both numbers would be rounded to the nearest half.

3. Determine if each statement is true or false. If the statement is false, give a counterexample.
   a. All rational numbers are integers.
   b. All fractions containing two integers are rational numbers.

Use estimation to simplify each expression.

4. $6.9 + 10.81$ 5. $0.07 + 0.8$ 6. $-0.13 + (-0.63)$

7. $3.24 - 9.17$ 8. $-1.7 - (-5.6)$ 9. $0.4 - (-0.573)$

10. $14.88 - 1.17$ 11. $37.1 - 6.74$ 12. $-0.7 + 0.4$

13. Minnie bought three presents for her mom for her birthday. One present cost $1.79, another present cost $4.05 and the third one cost $10.40. Find the approximate total cost of all of the birthday presents. Use words and/or numbers to show how you determined your answer.

14. Robert recorded the highest and lowest temperatures for the month of January in his hometown. The highest temperature was 45.3° F and the lowest temperature was −5.7°. Approximately how many degrees difference is there between the highest and the lowest temperature in his hometown?

15. Harris began a weight loss program and recorded his first four days of weight gains and losses: −0.4 *kg*, +0.12 *kg*, −0.9 *kg*, −0.61 *kg*. Estimate his total weight change over the last four days. Use words and/or numbers to show how you determined your answer.

Use estimation to simplify each expression.

16. $\frac{1}{8} + \frac{5}{11}$

17. $4\frac{1}{3} + 5\frac{5}{6}$

18. $9\frac{4}{5} + \left(-2\frac{7}{9}\right)$

19. $-\frac{3}{5} - \frac{2}{13}$

20. $\frac{7}{15} - \frac{5}{12}$

21. $10\frac{4}{11} - 13\frac{1}{6}$

22. $-1\frac{3}{20} + \left(-6\frac{9}{10}\right)$

23. $-\frac{11}{12} - \frac{4}{5}$

24. $5\frac{3}{16} - \left(-3\frac{2}{5}\right)$

25. Victor bought three items at a local farmer's market. He bought $2\frac{2}{3}$ pounds of tomatoes, $4\frac{1}{6}$ pounds of cucumbers and $5\frac{2}{7}$ pounds of corn. Approximate the total weight of his purchases. Show all work necessary to justify your estimate.

26. The table below shows the changes in the value of Happy Sands Towing stock over the last year. Estimate the total change in value for the stock last year.

Quarter	Change
1	$+3\frac{1}{5}$
2	$+6\frac{4}{15}$
3	$-4\frac{5}{6}$
4	$-2\frac{1}{10}$

27. Michelle's dog has been losing weight each week. Michelle estimates the total weight loss over the last three weeks at 4 pounds. Write three possible non-whole number amounts that could represent the dog's weight loss for each of the three weeks. Explain how you know your values fit this situation.

28. Francisco wanted to make sure he had enough money in his wallet to buy three items marked $4.13, $9.29 and $7.35. He rounded each amount to the nearest whole number and got a sum of $20, which is exactly what he had in his wallet. What is wrong with Francisco's estimation strategy? Explain your reasoning.

29. Write a real-world problem where four fraction or decimal amounts have a sum of approximately 10. Use mathematics to justify your answer.

REVIEW

Find each sum or difference. Write your answer in simplest form.

30. $\frac{5}{6} + \frac{2}{3}$

31. $4\frac{7}{10} - 2\frac{4}{5}$

32. $\frac{5}{12} + \frac{3}{8}$

33. $\frac{3}{4} + \frac{2}{5}$

34. $\frac{7}{9} - \frac{1}{3}$

35. $3\frac{1}{4} + 8\frac{1}{2}$

Find the value of each expression.

36. $5(2 - 6) + 4$

37. $\frac{2(2 + 3)^2}{5}$

38. $(-2 + 11)^2 - 8 \cdot 4$

Tic-Tac-Toe ~ Irrational Numbers

When you were younger, you learned about counting numbers in math class. As you have progressed in math you have learned about other classifications of numbers such as integers and rational numbers. Irrational numbers are another type of number you will work with in higher-level mathematics courses. Research irrational numbers to determine the following:

- How are irrational numbers different from rational numbers?
- When were irrational numbers first used?
- What are some common irrational numbers? What are they used for?
- Other interesting information about irrational numbers.

Write a 1–2 page paper that summarizes your findings. Cite all sources used.

Tic-Tac-Toe ~ Estimation Poetry

An acrostic poem is a special type of poem where the first letter of each line spells out a word. Write two acrostics using the word ESTIMATION. One acrostic poem should focus on situations where estimation of rational numbers is helpful. The other acrostic poem should be about situations where estimation of rational numbers may cause more harm than good.

ADDING RATIONAL NUMBERS

LESSON 3.2

 Add positive and negative fractions and decimals.

In **Block 1** of this book you reviewed adding fractions and decimals. You only worked with positive numbers at that time. The set of rational numbers includes positive fractions and decimals along with their opposites. In this lesson, you will learn how to add all rational numbers.

There are two common ways a negative fraction is written. The negative sign can be in front of the entire fraction or it can be found in the numerator. The negative may be in the denominator, but that is less common.

$$\text{Common forms: } -\frac{1}{4} \text{ or } \frac{-1}{4}$$

ADDING POSITIVE AND NEGATIVE FRACTIONS

1. Write all mixed numbers as improper fractions.
2. Rewrite fractions using the least common denominator. If fractions are negative, place the negative sign on the number in the numerator.
3. Use integer rules to add the numerators and write the sum over the common denominator.
4. Write the fraction in simplest form. If the answer is negative, place the negative sign in front of the sum.

EXAMPLE 1 Find the sum of $-\frac{5}{6} + \frac{1}{3}$.

SOLUTION Neither fraction is a mixed number so first rewrite the fractions with the least common denominator. The least common denominator is 6. Place the negative sign in the numerator.

$$\frac{-5}{6} \quad \text{and} \quad \frac{1}{3} \xrightarrow{\times 2} = \frac{2}{6}$$

Rewrite the problem with the numerators over the LCD. $\quad \frac{-5}{6} + \frac{2}{6}$

Use the integer rules to add the numerators and write the sum over the common denominator.

$$\frac{-5 + 2}{6} = \frac{-3}{6}$$

Zero pairs are removed.

Simplify the fraction and put the negative in front of the sum. $\frac{-3}{6} = -\frac{1}{2}$

$$-\frac{5}{6} + \frac{1}{3} = -\frac{1}{2}$$

EXAMPLE 2 The change in the value of Digimaxx stock last week was $-3\frac{1}{4}$ points. This week the change in the stock's value was $-1\frac{2}{3}$ points. Find the total change in value of Digimaxx stock over the last two weeks.

SOLUTION Find the sum of $-3\frac{1}{4} + \left(-1\frac{2}{3}\right)$.

Write each mixed number as an improper fraction as if the negative sign was not there. Write the negative sign in the numerator.

$$-3\frac{1}{4} = \frac{-13}{4} \qquad -1\frac{2}{3} = \frac{-5}{3}$$

The least common denominator is 12. Write equivalent fractions for each fraction with a denominator of 12.

$$\frac{-13}{4} = \frac{-39}{12} \qquad \frac{-5}{3} = \frac{-20}{12}$$

Add the numerators. Since both numerators are negative, the sum will be negative.

$$\frac{-39 + -20}{12} = \frac{-59}{12}$$

Write the sum over the common denominator and simplify. Put the negative sign in front of the mixed number.

$$\frac{-59}{12} = -4\frac{11}{12}$$

The total change in the value of Digimaxx stock was $-4\frac{11}{12}$ points.

ADDING POSITIVE AND NEGATIVE DECIMALS

1. Determine the sign of the answer.
2. Find the sum.
   - If the decimals have the same sign, add the absolute value of the numbers and use the sign determined in **Step 1**.
   - If the decimals have different signs, subtract the lesser absolute value from the greater absolute value and use the sign determined in **Step 1**.

Remember to use the integer sum rules when determining the sign of the answer.

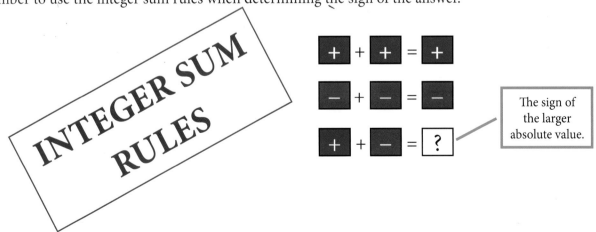

INTEGER SUM RULES

+ + + = +

− + − = −

+ + − = ?   The sign of the larger absolute value.

EXAMPLE 3 Find the value of −3.52 + 14.7.

SOLUTION Since one decimal is negative and the other is positive, the sum will have the sign of the number with the greater absolute value. In this case, the sum will be positive.

$$|-3.52| = 3.52$$
$$|14.7| = 14.7$$

The positive number has a greater absolute value so the sum will be positive.

Since the decimals have different signs, subtract the lesser absolute value from the greater absolute value. Line up the decimal points.

```
 14.70
− 3.52
 11.18
```

−3.52 + 14.7 = 11.18

EXERCISES

Find each sum. Write your answer in simplest form.

1. $-\frac{7}{10} + \left(-\frac{1}{10}\right)$

2. $-\frac{1}{8} + \frac{3}{8}$

3. $\frac{3}{8} + \left(-\frac{1}{3}\right)$

4. $-3\frac{5}{6} + \left(-5\frac{1}{2}\right)$

5. $1\frac{1}{3} + 4\frac{1}{2}$

6. $2\frac{3}{4} + \left(-7\frac{1}{2}\right)$

7. $2\frac{4}{5} + \left(-1\frac{1}{10}\right)$

8. $-\frac{1}{8} + \frac{3}{4}$

9. $-1\frac{1}{3} + \left(-2\frac{5}{6}\right)$

10. Emma's parents applied for a home loan which has an interest rate of $6\frac{1}{2}$%. Emma's parents have good credit, so the bank is willing to change their interest rate on their loan by $-1\frac{1}{4}$%. What will be the interest rate on Emma's parents' home?

11. A hedge has grown $\frac{7}{12}$ foot over the summer. Silas trimmed $\frac{3}{4}$ foot off the hedge.
   a. One of the fractions should be written as a negative number. Which one is it and why?
   b. Find the sum of the two numbers to determine the total change in the hedge's height.

12. Jonah went scuba diving. On his first dive he descended $9\frac{1}{3}$ feet, paused and then descended an additional $12\frac{5}{6}$ feet.
   a. Explain why $-9\frac{1}{3} + \left(-12\frac{5}{6}\right)$ represents his total change in depth.
   b. Find his total change in depth.

Find each sum.

13. −2.6 + 4.1

14. −0.41 + (−3.5)

15. 7.37 + (−10.8)

16. 10.25 + 6.2

17. −45.2 + (−9.3)

18. −0.093 + 0.14

19. 2.07 + (−3.4)

20. −4 + 2.25

21. −15.7 + (−3.9)

22. Becky has borrowed money from two friends and she wants to pay off her debts. Her debt to one friend is represented by the number −$14.55. Her debt to her other friend is represented by the number −$8.79. What number represents her total debt?

23. Timothy's grade increased by 1.65 percentage points after he turned in a large project last week. He decided to take a break from homework because his grade had gone up. His grade dropped 2.4 percentage points after not doing homework for one week.
  a. One of the decimals should be written as a negative number. Which one is it and why?
  b. Find the sum of the two numbers to determine the total change in Timothy's grade over the past two weeks.

24. LaSarah started hiking from 32.4 meters below sea level. She climbed 61.5 meters in elevation during the hike. What number represents the highest elevation she reached?

25. Alicia incorrectly added 34.6 and −5.7 as shown below. Explain what she did wrong and find the correct sum.

$$\begin{array}{r} 34.6 \\ +-5.7 \\ \hline 40.3 \end{array}$$

26. Find the sum of −2.38, 5.6, $4\frac{1}{4}$ and $-1\frac{2}{5}$. Show all work necessary to justify your answer.

REVIEW

Order each set of integers from least to greatest.

27. −2, −8, −6, −5

28. 1, −4, −2, 6, 0

29. −8, −7, −1, −9, −4

Write each power in expanded form. Find each value.

30. 2^4

31. $(-4)^3$

32. $\left(\frac{1}{5}\right)^2$

Copy and complete each statement using <, > or =.

33. $(-3)^2$ ⬤ -3^2

34. -5^2 ⬤ -6^2

35. $\left(\frac{1}{2}\right)^2$ ⬤ $\left(\frac{1}{3}\right)^2$

SUBTRACTING RATIONAL NUMBERS

LESSON 3.3

 Subtract positive and negative fractions and decimals.

Oxygen is part of the air you breathe. When you breathe in oxygen, it is in the form of a gas. Oxygen can also be in two other forms. It can be a liquid or a solid. Oxygen is a solid until it reaches its melting point of −368.77° F and turns into a liquid. The boiling point is where oxygen turns from a liquid to a gas. This occurs at −297.4° F. For how many degrees will oxygen remain a liquid before it turns into a gas?

SOLID ⟶ LIQUID ⟶ GAS
Melting Point   Boiling Point
−368.77° F   −297.4° F

To find how far apart two numbers are, you must find the difference between the two numbers.

$$-297.4 - (-368.77)$$

When subtracting integers, turn the subtraction symbol into an addition symbol by adding the opposite.

$$-297.4 + 368.77$$

Follow the rules for adding rational numbers. Determine the sign of the answer. In this case the absolute value of the positive number is the largest, so the answer will be positive.

Since the decimals have different signs, subtract the lesser absolute value from the greater absolute value and use the sign determined above (positive).

$$\begin{array}{r} 368.77 \\ -297.40 \\ \hline 71.37 \end{array}$$

Oxygen remains a liquid from −368.77° F to −297.4° F. The number of degrees between the solid and gaseous form of oxygen is 71.37° F.

SUBTRACTING POSITIVE AND NEGATIVE RATIONAL NUMBERS

1. Turn the subtraction symbol into an addition symbol by adding the opposite.
2. Follow the rules for adding fractions and decimals to determine the sum.

---

84   *Lesson 3.3 ~ Subtracting Rational Numbers*

EXPLORE!   WHAT'S THE DIFFERENCE?

Find the difference in each situation.

Step 1: Yoko had $34.86 in her checking account. She wrote a check that left her with a balance of −$12.89. Find the value of the check she wrote.

Step 2: The neighborhood bank was offering savings accounts in January that earn $3\frac{2}{5}$% interest. In February, the interest rate dropped to $2\frac{9}{10}$%. What number represents the interest rate change from January to February?

Step 3: Margaret found out that the lowest temperature ever recorded in Vermont is −50.1° F. The highest temperature ever recorded in Vermont is 105.1° F. What is the difference between these two temperatures?

Step 4: R.J. dove into the ocean and reached an elevation of $-26\frac{1}{2}$ feet (the distance below sea level). On his second dive, he reached an elevation of $-31\frac{3}{4}$ feet. What was the difference between his two dives?

Step 5: What would you consider the most important thing to remember when adding positive and negative fractions or decimals? Why?

EXAMPLE 1   Find the value of $-6\frac{1}{3} - 2\frac{1}{2}$.

SOLUTION

Change the subtraction symbol to an addition symbol by adding the opposite.
$$-6\tfrac{1}{3} - 2\tfrac{1}{2} \rightarrow -6\tfrac{1}{3} + \left(-2\tfrac{1}{2}\right)$$

Convert each mixed number to an improper fraction.
$$-6\tfrac{1}{3} = \tfrac{-19}{3} \qquad -2\tfrac{1}{2} = \tfrac{-5}{2}$$

Create equivalent fractions using the least common denominator of 6.
$$\tfrac{-19}{3} = \tfrac{-38}{6} \qquad \tfrac{-5}{2} = \tfrac{-15}{6}$$

Rewrite the expression with equivalent fractions.
$$\tfrac{-38}{6} + \tfrac{-15}{6}$$

Add the numerators.
$$\tfrac{-38 + (-15)}{6} = \tfrac{-53}{6}$$

Simplify.
$$\tfrac{-53}{6} = -8\tfrac{5}{6}$$

$-6\tfrac{1}{3} - 2\tfrac{1}{2} = -8\tfrac{5}{6}$

EXAMPLE 2 Bromine has a boiling point of 58.8° C and a melting point of −7.2° C. What is the difference between the boiling point and the melting point?

SOLUTION

Write a subtraction expression to represent the situation. $58.8 − (−7.2)$

Add the opposite. $58.8 − (−7.2) \rightarrow 58.8 + 7.2$

Add the two numbers by lining up the decimals.

$$\begin{array}{r} 58.8 \\ + 7.2 \\ \hline 66.0 \end{array}$$

A zero on the end of a decimal can be dropped.

The difference between the boiling point and the melting point is 66° C.

EXERCISES

Find each difference. Write each answer in simplest form.

1. $\frac{3}{8} − \left(−\frac{1}{8}\right)$

2. $−\frac{1}{9} − \frac{2}{3}$

3. $7\frac{1}{2} − \left(−3\frac{1}{4}\right)$

4. $−8 − 5\frac{1}{4}$

5. $−\frac{3}{10} − \left(−\frac{1}{5}\right)$

6. $1\frac{4}{9} − 2$

7. $−1\frac{2}{3} − 3\frac{2}{5}$

8. $\frac{1}{4} − \left(−\frac{4}{5}\right)$

9. $2\frac{2}{9} − 2\frac{1}{3}$

10. Katie was hiking in Death Valley, California. At the beginning of the hike she was at an elevation of $−37\frac{1}{4}$ feet. At the halfway point of her hike she had reached an elevation of $−82\frac{1}{2}$ feet. What was her change in elevation? Show all work necessary to justify your answer.

11. Josh was learning a new game. After the first hour of playing the game, Josh had a score of $−7\frac{1}{2}$. After the second hour, his score had risen to $−1\frac{1}{8}$. How many points did Josh score in the second hour? Explain how you know your answer is correct.

Find each difference.

12. $7.52 − (−4.3)$

13. $−0.6 − 3.1$

14. $7.37 − (−10.8)$

15. $5.4 − 6.9$

16. $−15.8 − 8.2$

17. $−0.056 − (−0.27)$

18. $70.9 − (−3.4)$

19. $−8 − 2.76$

20. $−0.7 − (−0.12)$

21. Josiah's checking account was at $104.24 on Monday. One of his checks was cashed on Tuesday and his account went down to −$23.99. What was the value of the check that was cashed? Show all work necessary to justify your answer.

22. The boiling point of nitrogen is −320.42° F. The melting point is 25.58° F colder. What is the melting point of nitrogen? Show all work necessary to justify your answer.

REVIEW

Find each product or quotient. Write each answer in simplest form.

23. $\frac{2}{5} \cdot \frac{3}{7}$

24. $\frac{4}{9} \cdot \frac{3}{8}$

25. $\frac{1}{2} \div \frac{3}{14}$

26. Colin bought $13\frac{1}{3}$ pounds of cat food from the bulk food bins. He plans to feed his cats the same amount of food each day. He wants the food to last for 20 days. How much food should he put out for his cats each day? Show all work necessary to justify your answer.

Tic-Tac-Toe ~ Stocks

 A share of stock represents a share of ownership in a company. Choose three different companies you would like to "buy" stock in. You are given $300 to spend on stock for each company.

1. On the first day, record the value of one share of each company's stock. You may want to use the internet or newspaper to research the current values of each stock. Determine how many stocks you can purchase from each company with $300. Show all work.

2. For the next four business days, record the change in value of each stock at the end of the day. Record the new value of one stock and the value of your purchase based on the changes. Use a table like the one shown below. Remember that the number of stocks purchased remains the same throughout the whole activity.

Example: Digimaxx Stock   Stock Purchased:

Date	Daily Starting Value of One Stock	Change in Value	Ending Value of One Stock	Current Worth of Stocks You Own in this Company

3. Find the total change in value of each purchase at the end of one week. Which stock did the best? Which stock did the worst?

4. Would you be interested in investing in stocks when you are an adult? Why or why not?

NOTE: This is a stocks simulation. You should not purchase any stocks while completing this activity.

ESTIMATING PRODUCTS AND QUOTIENTS

LESSON 3.4

 Estimate products and quotients of rational number operations.

Estimating products and quotients requires the use of **compatible numbers**. Compatible numbers are numbers that are easy to mentally compute.

EXPLORE!   IN YOUR HEAD

To get accurate answers for many rational number operations, you must have time, a pencil and paper to find the sum, difference, product or quotient. When you are in real-world situations, you may not have these things available to you. Compatible numbers are very important when estimating products and quotients.

Step 1: Jared's school gets $\frac{1}{5}$ of the sales from a coupon book fund raiser. He sold $55.75 worth of coupon books. He wants to determine the approximate amount of money his school gets from his sales.
  a. What calculation could Jared do to find the exact answer?
  b. Jared estimated the answer in his head. He changed his sales amount to a whole number. Which number, 55 or 56, is more compatible with $\frac{1}{5}$? Why?
  c. Use this number to find the approximate amount of money his school gets from his sales.

Step 2: Holly's dog, Spot, is on a strict weight-loss plan. Spot's change in weight over the last 4 weeks has been −35.2 ounces. Spot's weight loss has been steady over the past four weeks. Holly wants to determine what number represents Spot's approximate weight loss per week.
  a. What calculation could Holly do to find the exact answer?
  b. Holly estimated the answer in her head. She changed the ounces to a whole number. Which number, 34, 35 or 36, is more compatible with 4? Why?
  c. Use this number to determine Spot's approximate weight loss per week.

Step 3: Ronnie swam $43\frac{1}{4}$ miles over the past 7 weeks. She knows she swam the exact same amount each week. She wants to determine the approximate number of miles she is swimming each week.
  a. What calculation could Ronnie do to find the exact answer?
  b. Ronnie estimated the answer in her head. She changed her total distance to a whole number. The number is compatible with 7. What number did Ronnie change $43\frac{1}{4}$ to?
  c. Use this number to determine the approximate distance Ronnie swims each week. Which of the following would be the best way to phrase your answer and why?
     "She swam a little more than ____ miles each week."
     "She swam a little less than ____ miles each week."

EXPLORE! (CONTINUED)

Step 4: Estimate each of the following using compatible numbers. Explain each number change. Why did you choose those numbers for each estimation?

    a. $\frac{1}{3}(39.9)$ **b.** $29\frac{1}{3} \div 5$ **c.** $77\frac{1}{2} \div 10\frac{1}{5}$

Step 5: Give a situation in everyday life where compatible numbers may be used to multiply or divide. Write an example using that situation and estimate the solution by using compatible numbers.

ESTIMATING USING COMPATIBLE NUMBERS

1. Substitute compatible numbers for one or more rational numbers in the expression.
2. Find the value of the expression using the compatible number(s).

Not every rational number must be changed in an expression when estimating with products and quotients. In most situations mixed numbers should be changed to a compatible number while proper fractions can be left as is.

EXAMPLE 1

Estimate the product or quotient using compatible numbers.
a. $51.2 \div 6.94$ b. $-\frac{1}{4}\left(23\frac{2}{7}\right)$

SOLUTIONS

a. Write the expression. $51.2 \div 6.94$

Round the divisor to the nearest whole number. $51.2 \div 7$

Change the dividend to the nearest multiple of the new divisor and divide. $49 \div 7 = 7$

$51.2 \div 6.94 \approx 7$

b. Write the expression. $-\frac{1}{4}\left(23\frac{2}{7}\right)$

Change the second factor to the nearest multiple of four and multiply. $-\frac{1}{4}(24) = -6$

$-\frac{1}{4}\left(23\frac{2}{7}\right) \approx -6$

Lesson 3.4 ~ Estimating Products and Quotients

EXERCISES

1. Estimating is useful in many situations.
   a. Describe one situation where you would use estimation rather than determining the exact answer.
   b. Describe one situation where you would NOT use estimation and only an exact answer would be appropriate.

2. In your own words, define "compatible number".

Estimate each product or quotient using compatible numbers. Show all work necessary to justify your estimate.

3. $17.4 \div 3$

4. $-\frac{1}{8}\left(-23\frac{1}{4}\right)$

5. $0.25(41.2)$

6. $143.5 \div 20.4$

7. $53\frac{7}{8} \div 5\frac{1}{4}$

8. $-\frac{1}{7}(-61)$

9. $49.6 \div (-10.4)$

10. $83\frac{1}{4} \div 2$

11. $\frac{2}{3}(9.4)$

12. Melinda hit 35 pitches in batting practice. One-sixth of the hits were fly balls. About how many hits were fly balls? Explain your estimation strategy.

13. Chauncey withdrew $43.75 from his bank. He spent $3.80 per day. Approximately how many days did his money last? Explain your estimation strategy.

14. Lee is making a batch of fudge that calls for $2\frac{1}{4}$ cups of sugar. He wants to make $5\frac{3}{4}$ batches of fudge. Estimate how much sugar he needs.

15. LaTasha filled planters with potting soil. She had 36.3 cubic feet of potting soil to fill 7 planters. About how much potting soil did she put in each planter?

16. Carson was a chef at a summer camp. He had $22\frac{1}{4}$ gallons of milk in the refrigerator. He planned to use about $4\frac{3}{8}$ gallons per day. Approximately how many days will it be until he runs out of milk?

17. Sebastian spent $40.50 in $5\frac{1}{2}$ days. His parents wanted to know approximately how much he spent each day. His mother calculated $40 ÷ 5. She thought he spent about $8 per day. His father estimated Sebastian spent about $7 per day by calculating $42 ÷ 6. Both used compatible numbers. Which answer do you believe is more accurate? Explain your reasoning.

18. Delaney needed to estimate $-125.8 \div 8.7$. She rounded the first number to the nearest hundred and the divisor to the nearest 10 to get $-100 \div 10$. Did she use a good estimation strategy? Explain your reasoning.

19. Each goat on the Flying R Ranch ate approximately 1.3 pounds of grain each day. Farmer Bill wanted to buy enough grain for the next 3 months (91 days) for his 26 goats. About how much grain should he buy? Explain your estimation strategy.

REVIEW

Find the value of each integer expression.

20. 50(–4)   **21.** –32 ÷ (–8)   **22.** 22 + (–3)

23. –9 – 2   **24.** –12(–3)   **25.** –5 + (–5)

26. 15 – 18   **27.** –100 ÷ (–25)   **28.** 3(–15)

29. 3 + 43   **30.** –6 – (–6)   **31.** –4(–7)

Tic-Tac-Toe ~ Estimation Worksheet

Estimation is used in many real-life situations. Think of a variety of situations where you have used estimation or you think estimation might have been useful. Ask friends or family for additional situations where they may have used estimation. Use these situations to create a worksheet that has 20 real-life questions where estimation could be used. Type or clearly print each question. Include an answer key that has both the best estimate and the exact answer.

Tic-Tac-Toe ~ Chemical Elements

Everything in the world around us is made of chemical elements. For example, water is made up of a combination of hydrogen and oxygen. Diamonds are made of carbon. Choose four different chemical elements and find the following information about each element:

1. the name of the element

2. three interesting facts about the element

3. the melting point and boiling point of the element in degrees Fahrenheit

4. the difference between the melting point and the boiling point

5. at least one thing in nature that contains this element

Create a poster, write a paper or create your own way to display the information.

MULTIPLYING RATIONAL NUMBERS

LESSON 3.5

 Find products of positive and negative fractions and decimals.

When multiplying positive and negative rational numbers, you can determine the sign of your product before multiplying. Use the integer product rules to determine the sign of the answer.

INTEGER PRODUCT RULES

$+ \cdot + = +$

$- \cdot - = +$

$+ \cdot - = -$

$- \cdot + = -$

MULTIPLYING POSITIVE AND NEGATIVE RATIONAL NUMBERS

1. Determine the sign of the product.
   - The product of two numbers with the same sign is positive.
   - The product of two numbers with different signs is negative.
2. Find the product. If working with fractions, put any negative signs in the numerators while finding the product.

EXAMPLE 1 Find the value of $\frac{4}{9}\left(-\frac{3}{10}\right)$. Write in simplest form.

Determining the sign first can help you check your work after you find the product.

SOLUTION The numbers have different signs, so the answer will be negative.

Multiply. $\frac{4}{9}\left(-\frac{3}{10}\right) = \frac{4}{9} \cdot \frac{-3}{10} = \frac{-12}{90}$

Simplify. $\frac{-12}{90} \xrightarrow{\div 6} \frac{-2}{15}$

$\frac{4}{9}\left(-\frac{3}{10}\right) = \frac{-2}{15}$ or $-\frac{2}{15}$

92 Lesson 3.5 ~ Multiplying Rational Numbers

EXAMPLE 2 Find the value of $-3\frac{1}{5}\left(-2\frac{2}{3}\right)$. Write in simplest form.

SOLUTION The numbers have the same signs, so the answer will be positive.

Write each mixed number as an improper fraction. Write the negative signs in the numerators.

$$-3\frac{1}{5} = \frac{-16}{5} \quad \text{and} \quad -2\frac{2}{3} = \frac{-8}{3}$$

Multiply.

$$\frac{-16}{5}\left(\frac{-8}{3}\right) = \frac{-16}{5} \cdot \frac{-8}{3} = \frac{128}{15}$$

Simplify.

$$\frac{128}{15} = 8\frac{8}{15}$$

$$-3\frac{1}{5}\left(-2\frac{2}{3}\right) = 8\frac{8}{15}$$

EXAMPLE 3 The water level of Renee's swimming pool was dropping 3.4 centimeters every hour. What number represents the change in the water's depth after 0.25 hours?

SOLUTION Write an expression to represent the situation. $-3.4(0.25)$

The numbers have different signs, so the answer will be negative. Ignore the signs while multiplying. Make the final answer negative.

It is easiest to put the number with the most digits on the top.

Multiply the two numbers.

```
   0.25
 × 3.4
  0100
  075
  0850
```

Count the number of places after the decimal point in the two factors. Move the decimal this many places in the product. Start at the right and move left.   0.850

Zeros at the end of a decimal do not need to be written.

Write with the sign determined above.   −0.85

The pool's change in depth in 0.25 hours was −0.85 centimeters.

Lesson 3.5 ~ Multiplying Rational Numbers

EXERCISES

1. Give one reason why it might be important to determine the sign of the product before multiplying. Explain your reasoning.

Find each product. Write in simplest form.

2. $-\frac{1}{2}\left(-\frac{1}{4}\right)$

3. $-\frac{4}{5} \cdot \frac{3}{10}$

4. $\frac{3}{4}\left(-\frac{1}{3}\right)$

5. $-3\frac{3}{4}\left(-1\frac{2}{3}\right)$

6. $-4\left(2\frac{1}{8}\right)$

7. $\frac{1}{2}\left(-5\frac{1}{2}\right)$

8. Roy ran the ball on each of the last 3 plays of Friday night's football game. On each play he lost $5\frac{1}{4}$ yards.
   a. Which value in this situation should be represented by a negative number? Why?
   b. Write a multiplication expression to determine the number that represents the total change in yardage over the last 3 plays.
   c. Find the value of your expression from **part b**.

9. Write two fraction problems where the fractions multiply to be a positive number. One of your problems must include at least one negative number. Show all work necessary to justify your answer.

10. Vanessa's family vacationed at Jackson Hole at the end of the ski season. The family stayed at the lodge for $2\frac{1}{2}$ days. During that time, the snow level decreased $\frac{3}{8}$ foot each day.
    a. Which value in this situation should be represented by a negative number? Why?
    b. What number represents the total change in snow depth over the past $2\frac{1}{2}$ days? Use mathematics to justify your answer.

Find each product.

11. $5.2(-0.6)$

12. $-10.1(-2)$

13. $-0.7(-0.1)$

14. $-14.9(1.2)$

15. $1.25(2.5)$

16. $-9(-3.3)$

17. During a drought, the water level in a pond decreased 2.6 centimeters each week. This continued for 4.5 weeks.
    a. Which value in this situation should be represented by a negative number? Why?
    b. Write a multiplication expression to determine the number that represents the total change in water depth over the past 4.5 weeks.
    c. Find the value of your expression from **part b**.

18. Alan used a meal card in the university cafeteria to pay for his lunch. Each day he used his card for lunch, the balance remaining on the card decreased by $3.62. What number represents the total change in value on his card if he used the card for 8 lunches last month?

19. Michiko grew 3.1 *cm* each of the last two years and 4.5 *cm* each of the two years previous to that. What was her change in height over the past four years? Show all work necessary to justify your answer.

20. Bryan lost weight by walking every day. He lost an average of 1.6 kilograms each week for 7 weeks. What number represents his total change in weight?

21. Tim needs to multiply three numbers: 4.2, 5 and 2. Kyle told him that he should multiply 5 and 2 first and then multiply that by 4.2. Tim is pretty sure he needs to multiply numbers in the order they are listed. Whose strategy do you like better? Explain your reasoning.

22. Kaylin bought $2\frac{3}{4}$ pounds of carrots for $0.84 per pound. She also bought $1\frac{1}{2}$ pounds of bananas for $0.70 per pound. She handed the cashier $5.00. How much change did she receive? Show all work necessary to justify your answer.

REVIEW

Find each sum or difference. Write each answer in simplest form.

23. $-\frac{3}{8} + \frac{1}{4}$

24. $3\frac{1}{2} - 4\frac{1}{3}$

25. $10.3 + 0.7$

26. $\frac{7}{10} - \frac{1}{5}$

27. $0.38 - 1.45$

28. $-3.92 + (-4.1)$

29. Ahn wanted to purchase six items at the grocery store. He needed to spend less than $30. The items and their prices are listed in the table at right. Do you think Ahn can buy all of the items? Explain your reasoning.

ITEMS	
Salad	$3.29
Chips	$2.39
Juice	$3.79
Steak	$9.19
Cake	$7.05
Flour	$5.35

TIC-TAC-TOE ~ CLASSIFICATIONS OF NUMBERS

There are many different classifications of numbers. Research each different classification below. Make a poster that will help other students understand the relationship between the different types of numbers. Include examples or definitions when necessary.

Types of numbers:
- Integers
- Natural Numbers
- Rational Numbers
- Irrational Numbers
- Whole Numbers
- Real Numbers

TIC-TAC-TOE ~ SCIENTIFIC NOTATION

Scientific notation is a method used by scientists and mathematicians to express very large and very small numbers. Scientific notation is an exponential expression using a power of 10.

$$N \times 10^P$$

Use the following process to convert a large or small number into scientific notation:

Step 1: Locate the decimal point and move it left or right so there is only one non-zero digit to its left. This number represents the value of N.

Step 2: Count the number of places that you moved the decimal point in **Step 1**. This number represents the value of P. If you moved the decimal point to the left, the sign of P is positive. If you moved the decimal point to the right, the sign of P is negative.

For example:

A. Convert 52,000 to scientific notation.
  Step 1: Move the decimal point to the **left** so there is only one non-zero digit to its left.
  $52,000 \rightarrow 5.2$
  Step 2: Count how many places the decimal point was moved in the number above.
  5 2 0 0 0.
  The decimal point was moved 4 places to the **left**. Since the decimal point was moved left, the P value is positive. Scientific notation for 52,000 is 5.2×10^4.

B. Convert 0.00492 to scientific notation.
  Step 1: Move the decimal point to the **right** so there is only one non-zero digit to its left.
  $0.00492 \rightarrow 4.92$
  Step 2: Count how many places the decimal point was moved in the number above.
  0.0 0 4 9 2
  The decimal point was moved 3 places to the **right**. Since the decimal point was moved right, the P value is negative. Scientific notation for 0.00492 is 4.92×10^{-3}.

Write each large or small number in scientific notation.

1. 0.000058

2. 9,700

3. 2,000,000,000

4. 0.00921

5. 0.0000007

6. 81,400,000

7. 560

8. 0.001092

9. 750,000

10. 1,000,000,000,000

11. 0.0000021

12. 0.82

13. Find a very large number that represents something in nature. Explain what the number represents and write this number in standard notation and scientific notation.

14. Find a very small number that represents something in nature. Explain what the number represents and write this number in standard notation and scientific notation.

DIVIDING RATIONAL NUMBERS

LESSON 3.6

 Find quotients of positive and negative fractions and decimals.

On the first day in January, Snowy Peaks Resort reported an increase of $2\frac{1}{4}$ inches in the snow base. If this were to continue each day, how long would it take for the total increase in the snow base to equal 50 inches?

To determine how many days it will take for the snow base to increase 50 inches, divide the total amount of snow needed by the change per day. $50 \div 2\frac{1}{4}$

Write each number as an improper fraction. $50 = \frac{50}{1}$ and $2\frac{1}{4} = \frac{9}{4}$

Multiply by the reciprocal of the divisor. $\frac{50}{1} \div \frac{9}{4} = \frac{50}{1} \cdot \frac{4}{9} = \frac{200}{9}$

Simplify. $\frac{200}{9} = 22\frac{2}{9}$

If the snow continued to fall at this rate, it would increase 50 inches in $22\frac{2}{9}$ days.

DIVIDING POSITIVE AND NEGATIVE RATIONAL NUMBERS
1. Determine the sign of the quotient.
   - The quotient of two numbers with the same sign is positive.
   - The quotient of two numbers with different signs is negative.
2. Find the quotient as if working with positive numbers. Then put the sign on the quotient.

EXAMPLE 1 Find the quotient of $-5\frac{2}{3} \div (-4)$.

SOLUTION

The rational numbers have the same signs, so the quotient will be positive. Ignore signs until the quotient is determined.

$\boxed{-} \div \boxed{-} = \boxed{+}$

Write each number as an improper fraction. $5\frac{2}{3} = \frac{17}{3}$ and $4 = \frac{4}{1}$

Rewrite the expression with improper fractions. $\frac{17}{3} \div \frac{4}{1}$

Multiply by the reciprocal of the divisor. $\frac{17}{3} \cdot \frac{1}{4} = \frac{17}{12}$

Simplify. $\frac{17}{12} = 1\frac{5}{12}$

Put the sign on the quotient. $-5\frac{2}{3} \div (-4) = 1\frac{5}{12}$

Lesson 3.6 ~ Dividing Rational Numbers

EXAMPLE 2 Find the quotient of $9.36 \div (-5.2)$.

SOLUTION

The rational numbers have different signs, so the quotient will be negative. Ignore signs until the quotient is determined.

$+ \div - = -$

Move the decimal point in the divisor to the right to form a whole number.

$5.2 \rightarrow 52$

Move the decimal point in the dividend the same number of places.

$9.36 \rightarrow 93.6$

Divide the two numbers.

$$52 \overline{)93.6}$$
$$\underline{-52}$$
$$41\,6$$
$$\underline{-41\,6}$$
$$0$$

quotient: 1.8

Put the sign on the quotient. -1.8

$9.36 \div (-5.2) = -1.8$

EXAMPLE 3 Scott bought a bag of cake flour that contained $38\tfrac{1}{2}$ cups of flour. He makes muffins to sell at his bakery. Each batch of muffins requires $3\tfrac{1}{2}$ cups of flour. Determine how many batches of muffins he can make with this bag of flour.

SOLUTION

Divide the total amount of flour by the amount per batch.

$38\tfrac{1}{2} \div 3\tfrac{1}{2}$

The rational numbers have the same signs, so the quotient will be positive.

$+ \div + = +$

Write each number as an improper fraction.

$38\tfrac{1}{2} = \dfrac{77}{2}$ and $3\tfrac{1}{2} = \dfrac{7}{2}$

Multiply by the reciprocal of the divisor.

$\dfrac{77}{2} \div \dfrac{7}{2} = \dfrac{\cancel{77}^{11}}{\cancel{2}_1} \cdot \dfrac{\cancel{2}^1}{\cancel{7}_1} = \dfrac{11}{1}$

Scott will be able to make 11 batches of muffins with the bag of flour.

EXERCISES

1. Create an expression for each description. Use mathematics to justify your answer.
   a. Two fractions whose quotient is negative.
   b. Two decimals whose quotient is positive.

Find each quotient. Write in simplest form.

2. $\frac{2}{3} \div \left(-\frac{5}{7}\right)$

3. $-\frac{2}{5} \div \left(-\frac{7}{10}\right)$

4. $5 \div \left(-\frac{3}{4}\right)$

5. $-2\frac{1}{2} \div \left(1\frac{2}{3}\right)$

6. $-8 \div \left(-1\frac{1}{5}\right)$

7. $6\frac{1}{2} \div \left(-\frac{1}{2}\right)$

8. Eddie eats cereal with the same amount of milk every morning. Over the past 6 days, his gallon of milk has decreased by $4\frac{1}{2}$ cups.
   a. Which value in this situation should be represented by a negative number? Why?
   b. What number represents the change in volume of the gallon of milk? Show all work to support your answer.

9. A crew of construction workers worked a total of $31\frac{1}{2}$ hours on a project. Each crew member worked $3\frac{1}{2}$ hours on the project. How many members were part of the crew?

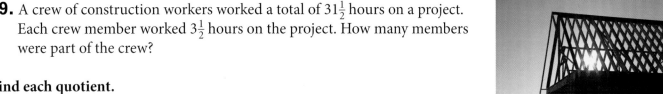

Find each quotient.

10. $15.4 \div (-1.1)$

11. $-44 \div (-0.25)$

12. $-0.7 \div 1.4$

13. $-165 \div (-7.5)$

14. $1.25 \div 0.1$

15. $11.05 \div (-2.6)$

16. Kristy has a gift card for the local movie rental store where all rentals are the same price. The value on her card decreased by $28.80 after 8 movie rentals.
   a. Which value in this situation should be represented by a negative number? Why?
   b. Write a division expression to determine the number that represents the change in value on her card for one movie.
   c. Find the value of your expression from **part b**.

17. Carlos made a cup of hot chocolate and left it sitting on the table. Each minute, the temperature of the hot chocolate decreased by 0.8° F. After awhile the temperature of the hot chocolate had decreased by 14.4° F.
   a. Which value(s) in this situation should be represented by a negative number? Why?
   b. How many minutes has Carlos' hot chocolate been cooling? Show all work necessary to support your answer.

18. Ryan bought flowers and chocolates for his parents for their anniversary. The flowers cost $48.95 and the chocolates cost $8.50. He and his two brothers split the cost of the items. How much did each brother pay? Show all work necessary to justify your answer.

19. The balance in Jenna's bank account has decreased by $170 over the last 8 weeks. She takes the exact same amount of money out of her account each week.
   a. What number represents the change in value of Jenna's bank account each week?
   b. Over the next three weeks she added $74 per week. What was her total change in value in the account over the eleven-week period? Use mathematics to justify your answer.

20. Tamira found the value of $4\frac{1}{3} \div 6\frac{2}{3}$ to be $\frac{13}{20}$. She wants to check her answer but does not just want to do the division problem over again. What is another way for her to check her answer? Is her answer correct? Support your answer using a method other than division.

REVIEW

Find the value of each expression. Show all work.

21. $\frac{12 + 4}{6 - 4} - 10$

22. $(-3 - 2)^2 + 8$

23. $-8(3 - 7) + 1$

24. $8 - \frac{1}{2}(-5 - 1)$

25. $\frac{3 \cdot 8 - 10}{5 - (-2)}$

26. $(1 - 3)^2 + (-1)^3$

27. Four friends order a pizza for $17, a two-liter bottle of soda for $2 and a tub of cookie dough for $5. They evenly split the cost. How much does each friend owe? Support your answer using mathematics.

Tic-Tac-Toe ~ Order of Operations

The order of operations is used to find the value of all types of numerical expressions. In **Block 2** you learned the order of operations and practiced it using integers. In this activity, you will use order of operations to find the value of a variety of expressions including fractions and decimals. Show all work and write each answer in simplest form.

1. $6.1(3.7 + 4.2) + 8.75$

2. $\frac{4}{7 - 2} + \frac{3}{4}$

3. $\frac{1}{2}(3\frac{2}{3} + 1\frac{1}{6}) + 4\frac{1}{3}$

4. $\frac{0.25 - 0.55}{0.4 - 0.3} + 0.25(4)$

5. $-\frac{1}{6} + (\frac{1}{3} + \frac{1}{2})^2$

6. $-1.6 + 4.5(3) - 7.2$

7. $10(3.4 - 5.5)^2$

8. $-3(5\frac{1}{2} - 7\frac{2}{5}) - 1\frac{4}{5}$

9. $\frac{6 - 10}{3} + (\frac{2}{3})^2$

REVIEW   BLOCK 3

Vocabulary

compatible numbers
rational numbers

Estimate sums and differences of rational numbers.
Add positive and negative fractions and decimals.
Subtract positive and negative fractions and decimals.
Estimate products and quotients of rational number operations.
Find products of positive and negative fractions and decimals.
Find quotients of positive and negative fractions and decimals.

Lesson 3.1 ~ Estimating Sums and Differences

Estimate each decimal expression.

1. 7.85 + 12.1
2. 0.03 + 0.94
3. −0.94 + (−0.58)
4. 8.21 − 9.09
5. −0.52 − (−0.43)
6. 5.4 − (−6.573)

7. Nicolette bought three presents for her dad for his birthday. The prices of the three presents were $4.79, $2.05 and $12.31. Approximate the total cost of all of the birthday presents.

Estimate each fraction expression.

8. $\frac{1}{10} + \frac{11}{13}$
9. $4\frac{1}{6} + 7\frac{7}{8}$
10. $11\frac{4}{5} + \left(-5\frac{7}{9}\right)$
11. $-\frac{3}{5} - \frac{1}{15}$
12. $\frac{9}{16} - \frac{7}{15}$
13. $6\frac{11}{12} - 10\frac{1}{6}$

14. Mark bought four items in the produce section at the grocery store. The items included $1\frac{5}{6}$ pounds of potatoes, $3\frac{1}{8}$ pounds of oranges, $2\frac{2}{7}$ pounds of broccoli and $\frac{11}{12}$ pound of beans. What was the approximate total weight of his purchase? Explain your estimation strategy.

Block 3 ~ Review 101

Lesson 3.2 ~ Adding Rational Numbers

Find each sum. Write your answer in simplest form.

15. $-\frac{2}{5} + \left(-\frac{1}{10}\right)$

16. $-\frac{1}{3} + \frac{4}{9}$

17. $-0.21 + (-6.3)$

18. $-4\frac{1}{2} + 1\frac{1}{2}$

19. $\frac{5}{12} + \left(-\frac{1}{4}\right)$

20. $0.92 + (-4.7)$

21. $-5.26 + (-14.8)$

22. $-4\frac{2}{3} + \left(-3\frac{1}{2}\right)$

23. $0.43 + (-0.42)$

24. Gina purchased a certificate of deposit (CD) which has an interest rate of $6\frac{1}{2}$%. Three months later, the interest rate the bank was offering for new CDs changed by $-1\frac{3}{8}$%. What was the new rate the bank was offering?

25. Daniel had borrowed money from his parents. He wants to pay off his debts. The amount he owes his mom is represented by the number $-\$18.65$. His debt to his dad is represented by the number $-\$12.18$. What number represents his total debt? Show all work necessary to justify your answer.

Lesson 3.3 ~ Subtracting Rational Numbers

Find each difference. Write your answer in simplest form.

26. $\frac{3}{7} - \left(-\frac{5}{7}\right)$

27. $-\frac{1}{4} - \frac{1}{2}$

28. $0.5 - 2.1$

29. $8.03 - (-7.9)$

30. $2\frac{1}{5} - \left(-4\frac{1}{10}\right)$

31. $-1.04 - 1.8$

32. $-16 - (-20.5)$

33. $\frac{1}{8} - \frac{2}{3}$

34. $-5\frac{1}{2} - 3\frac{1}{4}$

35. Abby was learning a new game. After the first hour of playing the game, Abby's score was $-4\frac{1}{3}$. After the second hour, her score rose to $5\frac{2}{3}$. How many points did Abby score in the second hour?

36. The boiling point of argon is $-302.53°$ F. The melting point of argon is $-308.83°$ F. What is the difference between the boiling point and the melting point? Explain how you know your answer is correct.

Lesson 3.4 ~ Estimating Products and Quotients

Estimate each product or quotient using compatible numbers.

37. $22.3 \div 7$

38. $\frac{1}{5}\left(-36\frac{1}{4}\right)$

39. $26\frac{1}{7} \div 9$

40. $0.25(19.2)$

41. $103\frac{7}{8} \div 24\frac{1}{4}$

42. $-0.5(40.9)$

43. $-153.5 \div (-24.9)$

44. $43\frac{7}{8} \div 4\frac{1}{4}$

45. $-\frac{1}{8}(63)$

46. Larson hit 51 pitches in batting practice. About one-seventh of the hits were fly balls. About how many hits were fly balls? Explain your estimation strategy.

47. Katrina withdrew $67.75 from her bank. She spent $10.55 per day. Approximately how many days did her money last? Show all work necessary to justify your answer.

Lesson 3.5 ~ Multiplying Rational Numbers

Find each product. Write in simplest form.

48. $-\frac{3}{8}\left(-\frac{1}{2}\right)$

49. $-0.25(1.4)$

50. $-5.2(-2.8)$

51. $\frac{1}{5}\left(-\frac{5}{11}\right)$

52. $-2\frac{3}{5}\left(-1\frac{1}{3}\right)$

53. $0.06(-8)$

54. $5.2(-0.6)$

55. $\frac{5}{6}\left(-\frac{1}{4}\right)$

56. $5\left(6\frac{1}{2}\right)$

57. Josephine's family went on a vacation to the beach last summer. She watched the ocean tide go out at a rate of $1\frac{3}{10}$ feet per hour. The tide went out at this rate for $5\frac{1}{2}$ hours.
   a. Which value(s) in this situation should be represented by a negative number? Why?
   b. Write a multiplication expression to determine the number that represents the total change in the tide.
   c. Find the value of your expression from **part b**.

58. Bobby received a gift card for a local ice cream shop. Every day he used his card for an ice cream cone, the balance remaining on the card decreased by $2.65. What number represents the total change in value on his card if he used the card for 7 ice cream cones?

Lesson 3.6 ~ Dividing Rational Numbers

Find each quotient. Write in simplest form.

59. $\frac{7}{8} \div \left(-\frac{1}{2}\right)$

60. $-\frac{1}{4} \div \left(-\frac{5}{8}\right)$

61. $-29.7 \div (-9)$

62. $16.8 \div (-0.7)$

63. $9 \div \left(-1\frac{1}{4}\right)$

64. $-0.4 \div 1.6$

65. $\frac{3}{10} \div \left(-\frac{1}{15}\right)$

66. $-4\frac{2}{3} \div \left(-2\frac{1}{3}\right)$

67. $44.8 \div 0.8$

68. Eliza drinks a small glass of orange juice every morning. Over the past 8 days, her container of orange juice decreased by $5\frac{1}{3}$ cups.
    a. Which value in this situation should be represented by a negative number? Why?
    b. What number represents the daily change in the volume of orange juice in the container? Show all work to support your answer.

69. Jeremie's family returned to a cold house after being gone for the holidays. The temperature in the house increased by 0.6° F each minute after they turned the heat on. The furnace went off after the temperature in the house increased a total of 14.7° F. How many minutes was the heat on in Jeremie's house? Use words and/or numbers to show how you determined your answer.

70. Ivanka had $98.50. She spent $37.74 on a pair of jeans. She split the remaining money with her sister. How much does she have left? Show all work necessary to justify your answer.

Tic-Tac-Toe ~ Tutorial

Many students move from one school to another during the school year. Some students may enter your school without knowing how to add and subtract rational numbers including fractions and decimals. Create a tutorial booklet that students could use to catch up with their classmates. Include examples for each type of operation.

Create a one-page quiz that students can take to determine if they have learned the material after studying the tutorial booklet. Include an answer key on a separate sheet of paper.

CAREER FOCUS

DAVID
ELECTRONICS TECHNICIAN

My name is David and I am an Electronics Technician. My main responsibility is maintaining my district's radio equipment. We use portable hand-held radios. Our vehicles have radios in them as well. Radios allow us to communicate with each other and other agencies. I install radios, program them, fix problems and make repairs. I also install and maintain cell phone systems, satellite phones, and other electrical devices that one might find in a fire truck or law enforcement vehicle. In addition to these responsibilities, I maintain our building's phone system, manage my budget, order parts, keep track of inventory and work with many different computer applications.

I use mathematical formulas when fixing and maintaining electronics on a daily basis. I often need to calculate things like voltage drops, current draw, and power consumption. I also use math to calculate antenna lengths or power losses in cables. Work involving electronics requires using some type of math most of the time.

I started my career with six years in the Air Force as an Aviation Electronics Technician. I took many job-related courses during that time and gained priceless experience. After leaving the Air Force, I went to college and acquired a degree in electronics. This degree required many hours of study in many courses, including math. Most electronics-related jobs require on-going training involving high-level math.

An electronics technician earns between $40,000 - $70,000 or more a year. There are also electronic engineering jobs with salaries much higher than that.

I love my job. It is something I enjoy doing day after day and year after year. It is also challenging and rewarding. I like having a career that gives me opportunities to grow and advance. Most importantly, I like my job because it is a career where my efforts make a difference in people's lives.

CORE FOCUS ON RATIONAL NUMBERS & EQUATIONS

BLOCK 4 ~ SOLVING EQUATIONS

Lesson 4.1	Expressions and Equations	108
Lesson 4.2	Solving One-Step Equations	113
	Explore! Introduction to Equation Mats	
Lesson 4.3	Solving Two-Step Equations	118
	Explore! Equation Mats for Two-Step Equations	
Lesson 4.4	The Distributive Property	122
Lesson 4.5	Simplifying Expressions	126
	Explore! Where Do I Belong?	
Lesson 4.6	Simplifying and Solving Equations	130
	Explore! Equation Manipulation	
Lesson 4.7	Solving Equations with Variables on Both Sides	134
Lesson 4.8	Linear Inequalities	138
Review	Block 4 ~ Solving Equations	142

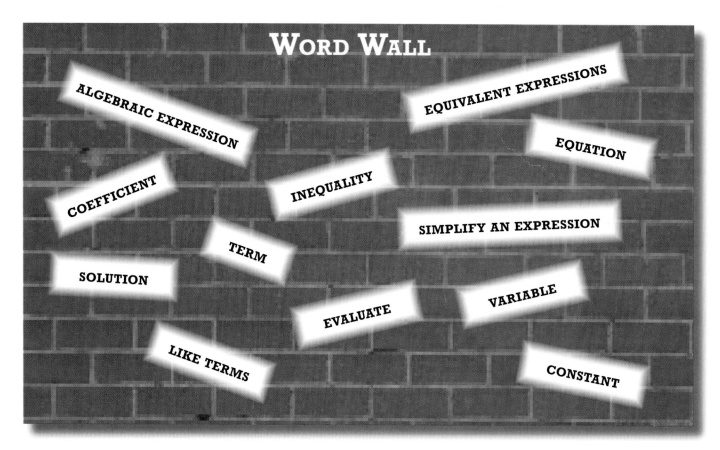

BLOCK 4 ~ SOLVING EQUATIONS
TIC-TAC-TOE

EQUATION POSTER Design a poster which explains how to solve multi-step equations. *See page 137 for details.*	**GIVEN AN ANSWER** Write two-step inequalities that have a given answer. *See page 145 for details.*	**BMI** Use the Body Mass Index formula in different situations. *See page 112 for details.*
NATE'S BUSINESS Examine Nate's business plan to help him determine profits and costs. *See page 129 for details.*	**FRACTION EQUATIONS** Solve real-life equations involving fractions. *See page 117 for details.*	**MATH DICTIONARY** Make a dictionary for all the vocabulary terms in this textbook. Create diagrams when possible. *See page 121 for details.*
ABSOLUTE VALUE Solve equations involving absolute value. Check your solutions. *See page 133 for details.*	**LETTER TO THE EDITOR** Write a letter to the editor about the graduation requirements in your state. *See page 125 for details.*	**EQUATION MATS** Produce a "How To..." guide to teach others how to use equation mats to solve equations. *See page 137 for details.*

EXPRESSIONS AND EQUATIONS

LESSON 4.1

Write and evaluate algebraic expressions.
Determine if a number is a solution of an equation.

Patty's mother owns a flower shop. On Mother's Day, Patty helps her mother deliver flowers. Her mother pays her $10 for the day plus an additional $2 for each delivery she makes. The expression that represents the total amount of money which Patty has made is:

$$10 + 2d$$

A **variable** is a symbol that represents one or more numbers. In this expression, d represents the number of deliveries Patty has made. An **algebraic expression** is a mathematical expression that contains numbers, operations (such as add, subtract, multiply or divide) and variables. You can **evaluate** an algebraic expression by substituting a number for the variable to find its value. Some algebraic expressions will have more than one variable. If this is the case, you will substitute each variable with a given number. After substituting each variable with a number, you will use the order of operations to find the value of the expression. The value of an algebraic expression changes depending on the value of the variable.

If Patty made twelve deliveries on Mother's Day, you can evaluate the expression by substituting 12 for d.

$$10 + 2(12) = \$34$$

Patti made $34 delivering flowers.

EVALUATING AN EXPRESSION

1. Rewrite the expression by substituting the given value(s) for the variable(s).
2. Follow the order of operations to find the value of the expression.

EXAMPLE 1 Evaluate each algebraic expression.
a. $5x - 3$ when $x = 8$
b. $(y + 2)^2 + p$ when $y = -7$ and $p = 9$

SOLUTIONS

a. Write the expression. $5x - 3$
Substitute 8 for x. $5(8) - 3$
Multiply. $40 - 3$
Subtract. 37

b. Write the expression. $(y + 2)^2 + p$
Substitute -7 for y and 9 for p. $(-7 + 2)^2 + 9$
Add inside parentheses. $(-5)^2 + 9$
Square -5. $25 + 9$
Add. 34

When you write algebraic expressions, it is necessary to know key words that represent the four basic operations in mathematics.

Addition	Subtraction	Multiplication	Division
sum	difference	product	quotient
increased by	decreased by	multiplied by	divided by
more than	less than	times	
plus	minus	of	

EXAMPLE 2

Write an algebraic expression for each phrase.
a. the product of seven and m
b. two more than six times x
c. f divided by three then decreased by four

SOLUTIONS

a. "product" means multiplication $\quad\quad\quad\quad 7m$

b. "more than" means addition
"times" means multiplication $\quad\quad\quad\quad 2 + 6x$ or $6x + 2$

c. "divided by" means division $\quad\quad\quad\quad \dfrac{f}{3} - 4$
"decreased by" means subtraction

Once you are able to write and evaluate expressions, you are able to work with equations. An **equation** is a mathematical sentence that contains an equals sign (=) between two expressions. Equations that involve a variable are neither true nor false until the equation is evaluated with a given value for the variable. A value is considered the **solution** of an equation if it makes the equation true.

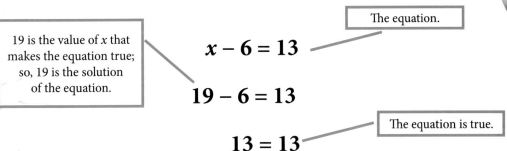

19 is the value of x that makes the equation true; so, 19 is the solution of the equation.

$x - 6 = 13$ — The equation.

$19 - 6 = 13$

$13 = 13$ — The equation is true.

To test if a value makes an equation true, substitute the value for the variable. If one side of the equation is equal to the other side, the value is a solution to the equation.

Lesson 4.1 ~ Expressions and Equations

EXAMPLE 3 Determine if the number given is the solution of the equation.
a. $-2y + 6 = -16$ Is 5 *the solution?*
b. $\frac{x}{5} - 4 = 4$ Is 40 *the solution?*

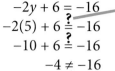

SOLUTIONS

a. See if the given value makes the equation true.
Write the equation. $-2y + 6 = -16$
Substitute 5 for y. $-2(5) + 6 \stackrel{?}{=} -16$
Multiply. $-10 + 6 \stackrel{?}{=} -16$
Add. $-4 \neq -16$

5 is NOT the solution.

b. Write the equation. $\frac{x}{5} - 4 = 4$

Substitute 40 for x. $\frac{40}{5} - 4 \stackrel{?}{=} 4$

Divide. $8 - 4 \stackrel{?}{=} 4$
Subtract. $4 = 4$

40 IS the solution.

EXERCISES

Write an algebraic expression for each phrase.

1. the sum of y and five

2. nine times m then increased by six

3. six less than the product of z and seven

4. the quotient of c and three

5. ten more than twice x

6. one subtracted from p

Write a phrase for each algebraic expression.

7. $y - 2$

8. $4d - 7$

9. $60 + 2x$

10. Write two different phrases for $3x + 2$. Underline the key words you used for the operations. Are there different key words you could have used? If so, give one more example.

Evaluate each expression.

11. $x - 5$ when $x = 24$

12. $12b + 3$ when $b = -4$

13. y^2 when $y = 5$

14. $50 - 7k$ when $k = 10$

15. $\frac{1}{2}m + \frac{3}{4}$ when $m = \frac{1}{4}$

16. $1.3h + 3.7$ when $h = 5$

110 *Lesson 4.1 ~ Expressions and Equations*

Evaluate each expression.

17. $-3d - 2c$ when $d = 2$ and $c = 10$

18. $\frac{2v}{5} - 3w$ when $v = 40$ and $w = 10$

19. Carol Middle School is planning a special "Movie Day" for all students who had no missing assignments last quarter. The principal will purchase enough containers of popcorn and juice for all the students that will be attending. The popcorn costs $2.50 per container. Bottles of juice cost $3.00 each. The principal uses the algebraic expression $2.50x + 3.00y$ to calculate the total expenses.
  a. What does the x variable represent?
  b. What does the y variable represent?
  c. The principal bought 8 containers of popcorn and 11 bottles of juice. How much did she spend? Show all work necessary to justify your answer.

20. The baseball team sold boxes of oranges and holiday wreaths as a fundraiser. The team made a profit of $6.25 for each box of oranges sold. The team made $8.00 in profit for each wreath sold.
  a. Let b represent a box of oranges and w represent a holiday wreath. Write an expression that represents the total profit made based on the number of boxes of oranges and wreaths that are sold.
  b. The team sold 40 boxes of oranges and 25 holiday wreaths. What was their total profit? Explain if your answer makes sense for the situation.

Determine if the number given is the solution of the equation. Use mathematics to justify your answer.

21. $2x - 6 = 3$ *Is 5 the solution?*

22. $-5y - 4 = -19$ *Is 3 the solution?*

23. $\frac{m}{8} + 10 = 12\frac{1}{2}$ *Is 20 the solution?*

24. $\frac{2}{3}h - 2 = -2$ *Is −6 the solution?*

25. $2(x + 4)^2 = 19$ *Is −1 the solution?*

26. $-2 + 10x = 13$ *Is 1.5 the solution?*

Write an algebraic equation for each phrase. Use x as the variable for the missing number. Match each equation to its solution. Use mathematics to justify your answer.

27. a number plus six equals ten   **A.** $x = 21$

28. three times a number then decreased by fourteen equals four   **B.** $x = 6$

29. a number divided by six then increased by four is nine   **C.** $x = 7$

30. a number squared is forty-nine   **D.** $x = 30$

31. nine more than half of a number is fourteen   **E.** $x = 4$

32. one more than the quotient of a number and three is eight   **F.** $x = 10$

REVIEW

Find each sum, difference, product or quotient.

33. $2\frac{1}{4}\left(-5\frac{1}{3}\right)$

34. $\frac{4}{5} + \left(-\frac{1}{2}\right)$

35. $1\frac{1}{2} - 3\frac{3}{4}$

36. $-\frac{8}{9} \div (-2)$

37. $5 - \left(-1\frac{2}{7}\right)$

38. $\frac{5}{27}\left(-\frac{9}{10}\right)$

Tic-Tac-Toe ~ BMI

Body Mass Index (BMI) is a number calculated from a person's weight and height. It is a reliable indicator of body fat for adults. BMI is also an inexpensive and easy-to-perform method of screening for weight categories that may lead to health problems. BMI is calculated using the formula:

$$\frac{\text{weight} \cdot 703}{(\text{height in inches})^2}$$

BMI scores can be used to categorize adults into standard weight status categories:

BMI	Weight Status
Below 18.5	Underweight
18.5 – 24.9	Normal
25.0 – 29.9	Overweight
30.0 and Above	Obese

Calculate the BMI for each individual and classify them using the weight status categories above. Show all work.

1. Female, 5' 5", 140 pounds

2. Male, 6' 1", 275 pounds

3. Male, 5' 10", 180 pounds

4. Female, 5' 7", 115 pounds

Write an equation and solve it to find the weight of a person with the given height and BMI. Show all work and round the answer to the nearest tenth of a pound.

5. Male, 5' 8", BMI = 22.2

6. Female, 5' 1", BMI = 30

7. Male, 6' 4", BMI = 27.4

8. Female, 5' 9", BMI = 18.3

9. Research obesity to find at least three diseases that are caused by obesity.

10. Using complete sentences, write out at least two ways people can help keep their weight status in the normal BMI range.

SOLVING ONE-STEP EQUATIONS

LESSON 4.2

  Use inverse operations to solve one-step equations.

Katie is three times as old as her cousin Maleah. If Katie is 14 years old, how old is Maleah?

An equation can be written to represent this situation. Let m represent Maleah's age.

$$3m = 14$$

To solve an equation, the variable must be isolated on one side of the equation. This process is sometimes referred to as "getting the variable by itself". In order to determine Maleah's age, you must get m by itself on one side of the equation. The most important thing to remember when solving an equation is that the equation must always remain balanced. Whatever operation occurs on one side of the equation must be performed on the other side to keep both sides equal.

THE PROPERTIES OF EQUALITY

For any numbers a, b and c:

Subtraction Property of Equality	If $a = b$, then $a - c = b - c$
Addition Property of Equality	If $a = b$, then $a + c = b + c$
Multiplication Property of Equality	If $a = b$, then $ac = bc$
Division Property of Equality	If $a = b$, then $\frac{a}{c} = \frac{b}{c}$

EXPLORE!   **INTRODUCTION TO EQUATION MATS**

Step 1: If you do not have an equation mat, draw one like the one seen below on a blank sheet of paper.

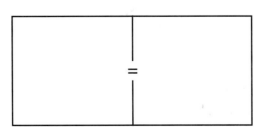

Lesson 4.2 ~ Solving One-Step Equations 113

EXPLORE! (CONTINUED)

Step 2: On your equation mat, place a variable cube on one side with 3 negative integer chips. On the other side of the mat, place 5 positive integer chips. This represents the equation $x - 3 = 5$.

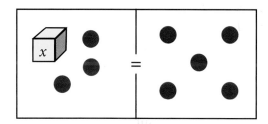

Step 3: In order to get the variable by itself, you must cancel out the 3 negative integer chips with the variable. Use zero pairs to remove the chips by adding three positive integer chips to the left side of the mat. Whatever you add on one side of the mat, add on the other side of the mat. This is using the Addition Property of Equality. How many chips are on the right side of the mat? What does this represent?

Step 4: Clear your mat and place chips and variable cubes on the mat to represent the equation $4x = -8$. Draw this on your own paper.

Step 5: Divide the integer chips equally among the variable cubes. This is using the Division Property of Equality. Each variable cube is equal to how many integer chips? Write your answer in the form $x = $ ___.

Step 6: Model an addition equation on your mat. Record the algebraic equation on your paper.

Step 7: Solve your equation. What does your variable equal? Explain how you know your answer is correct.

You will not always have integer chips or an equation mat available to use when you are solving equations. You can solve equations using inverse operations to keep your equation balanced. Inverse operations are operations that undo each other, such as addition and subtraction.

Operation	Inverse Operation
Addition	Subtraction
Subtraction	Addition
Multiplication	Division
Division	Multiplication

For example, if an equation has a number being added to the variable, you must subtract the number from both sides of the equation to perform the inverse operation. Even though you may be able to solve many one-step equations mentally, it is important that you show your work. The equations you will be solving in later lessons and in future math classes will become more complex.

EXAMPLE 1

Solve each equation. Show your work and check your solution.
a. $x + 13 = 41$ b. $6m = 27$ c. $-4 = \dfrac{y}{3}$

SOLUTIONS

Drawing a vertical line through the equals sign can help you stay organized. Whatever is done on one side of the line to cancel out a value must be done on the other side.

a. $x + 13 = 41$

The inverse operation of addition is subtraction.
Subtract 13 from both sides of the equation to isolate the variable.

$$\begin{array}{r|r} x + \cancel{13} & 41 \\ -\cancel{13} & -13 \\ \hline x = 28 & \end{array}$$

☑ Check the answer by substituting the solution into the original equation for the variable.

$28 + 13 \stackrel{?}{=} 41$
$41 = 41$

b. $6m = 27$

Divide both sides of the equation by 6.

$$\dfrac{\cancel{6}m}{\cancel{6}} \Big| \dfrac{27}{6}$$

$m = 4.5$ or $m = 4\dfrac{1}{2}$

☑ Check the solution.

$6(4.5) \stackrel{?}{=} 27$
$27 = 27$

c. $-4 = \dfrac{y}{3}$

The variable can be on either side of the equation.

Multiply both sides of the equation by 3.

$3 \cdot -4 \Big| \dfrac{y}{\cancel{3}} \cdot \cancel{3}$

$-12 = y$

☑ Check the solution.

$-4 \stackrel{?}{=} \dfrac{-12}{3}$
$-4 = -4$

EXAMPLE 2

The meteorologist on Channel 3 announced that a record high temperature had been set in Kirkland today. The new record is 2.8° F more than the old record. Today's high temperature was 98.3° F. What was the old record? Write a one-step equation and solve.

SOLUTION

Let x represent the old record temperature.
The equation which represents this situation is:

$x + 2.8 = 98.3$

Subtract 2.8 from each side of the equation to isolate the variable.

$$\begin{array}{r|r} x + \cancel{2.8} & 98.3 \\ -\cancel{2.8} & -2.8 \\ \hline x = 95.5 & \end{array}$$

☑ Check.

$95.5 + 2.8 \stackrel{?}{=} 98.3$
$98.3 = 98.3$

The old record high temperature in Kirkland was 95.5° F.

EXERCISES

Solve each equation using an equation mat or inverse operations. Show all work necessary to justify your answer.

1. $x - 6 = 18$
2. $y + 8 = -3$
3. $\frac{p}{3} = 11$
4. $3.4 = m - 7.7$
5. $9k = 72$
6. $x + \frac{2}{3} = \frac{5}{6}$
7. $20 = \frac{w}{-6}$
8. $-6d = -24$
9. $f + 0.8 = 2$
10. $-19 = b - 1$
11. $\frac{1}{4}x = 3$
12. $\frac{h}{12} = -0.5$
13. $6 = -a$
14. $g - 1.39 = 0.6$
15. $-3 + y = 10$

16. Reese did not check her solutions for the four-problem quiz on one-step equations. Check Reese's answers. If she got them incorrect, find the correct answer.

   a. $8x = 40$

   Reese's answer: $x = 32$

   b. $-2.6 + x = 4.9$

   Reese's answer: $x = 2.3$

   c. $6 = \frac{x}{-2}$

   Reese's answer: $x = -12$

   d. $x - \frac{1}{2} = 3\frac{1}{2}$

   Reese's answer: $x = 3$

17. Determine the age of Katie's cousin, Maleah, from the beginning of this lesson (page 113).

18. Students are selling wrapping paper for a school fundraiser. Keisha still needs to sell $24.50 worth of wrapping paper to reach her goal. She has already sold $48.50 worth of wrapping paper. Let w represent the dollar amount of wrapping paper Keisha wants to sell. What is the total value of wrapping paper Keisha needs to sell? Show all work necessary to justify your answer.

19. Willis is thinking of two numbers. Their sum is −5. If one of the numbers is 19, what is the other number? Write an equation and solve it to find the other number. Explain how you know your answer is correct.

20. In 2001, Roman Sebrle of the Czech Republic set a world record in the decathlon. In 2012, Ashton Eaton of the United States broke that record by scoring 9039 points. Eaton's score was 13 points more than Sebrle's record. How many points did Roman Sebrle score to set the world record? Use numbers, words and/or an equation to show how you determined your answer.

Write an algebraic equation for each sentence. Solve the equation using an equation mat or inverse operations.

21. The sum of x and twelve equals thirty-five.

22. The quotient of x and negative nine is negative two.

23. Twenty-two less than x equals negative six.

24. The product of x and one-half is five.

25. Explain in words how the Division Property of Equality helps solve a multiplication equation like $-4x = 44$.

REVIEW

Evaluate each expression where $x = 3$, $y = 9$ and $z = -4$.

26. $x + y + z$

27. $2(y - x)$

28. $\frac{y}{z}$

29. $11 + 2y - x$

30. $\frac{1}{2}(x + z) + y$

31. $-3z + \frac{2}{3}y$

Tic-Tac-Toe ~ Fraction Equations

Solving one-step multiplication equations with fractions requires the use of reciprocals. When solving multiplication equations you should use the inverse operation (which is division). If you are dividing by a fraction, multiply both sides of the equation by the reciprocal of the fraction.

For example: $\frac{2}{3}x = \frac{1}{4}$

$\frac{3}{2} \cdot \frac{2}{3}x = \frac{1}{4} \cdot \frac{3}{2}$ Multiply both sides by the reciprocal of $\frac{2}{3}$ which is $\frac{3}{2}$.

$x = \frac{3}{8}$

✓ $\frac{2}{3}\left(\frac{3}{8}\right) = \frac{1}{4}$

Solve each fraction equation. Write the answer in simplest form. Show all work. Check the answer.

1. $\frac{3}{5}x = \frac{1}{4}$

2. $\frac{1}{4}y = 6$

3. $\frac{4}{7}x = -\frac{4}{5}$

4. $-\frac{7}{10}m = 3$

5. $\frac{3}{4}p = 1\frac{2}{3}$

6. $\frac{1}{2}y = -5\frac{1}{4}$

7. $-\frac{7}{8}d = -2\frac{3}{8}$

8. $\frac{11}{12}h = 5$

9. $\frac{1}{5}x = 4\frac{7}{10}$

SOLVING TWO-STEP EQUATIONS

LESSON 4.3

Solve two-step equations.

Deborah opened an ice cream stand at the local pool last summer. She spent $148 on a freezer to start her business. She earned a profit of $0.75 for each ice cream cone sold. She earned a total of $452 during the summer.

Let x represent the number of cones that Deborah sold. The equation that models this situation is:

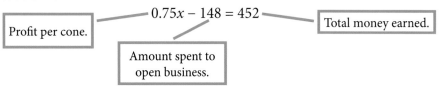

This equation is called a two-step equation. Two-step equations have two different operations in the equation. The two-step equation above has both multiplication and subtraction.

EXPLORE! **EQUATION MATS FOR TWO-STEP EQUATIONS**

Step 1: If you do not have an equation mat, draw one like the one seen below on a blank sheet of paper.

Step 2: On your equation mat, place two variable cubes on one side with 2 positive integer chips. On the other side of the mat, place 8 negative integer chips. What two-step equation does this represent?

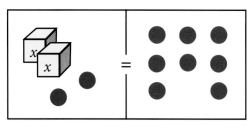

Step 3: The first step to solve an equation is to isolate the variables. The integer chips that are with the variables must be cancelled out. Use zero pairs to cancel out the two positive integer chips on the left side of the mat. Remember that whatever you add to one side must be added to the other side of the mat. Draw a picture of what your mat looks like now.

Step 4: Divide the chips on the mat equally between the two variable cubes. How many chips are equal to one cube? This is what x equals.

Step 5: Solve the equation $3x - 5 = 10$ using the equation mat. What does x equal?

118   Lesson 4.3 ~ Solving Two-Step Equations

In order to get the variable by itself in a two-step equation, you must perform two inverse operations. The inverse operations must undo the order of operations. That means you must start by undoing any addition or subtraction and then use inverse operations to remove any multiplication or division.

EXAMPLE 1

Using the equation at the beginning of this lesson, determine the number of ice cream cones Deborah sold last summer. Remember that x represents the number of ice cream cones sold. $0.75x - 148 = 452$

SOLUTION

Write the equation.
Add 148 to both sides.
Divide each side by 0.75

$$0.75x - 148 = 452$$
$$\underline{+148 \quad +148}$$
$$\frac{0.75x}{0.75} = \frac{600}{0.75}$$
$$\boxed{x = 800}$$

✓ Check the solution.

$$0.75(800) - 148 \stackrel{?}{=} 452$$
$$600 - 148 \stackrel{?}{=} 452$$
$$452 = 452$$

Deborah sold 800 ice cream cones last summer.

EXAMPLE 2

Solve the equation for the variable. Check the solution.
$$\frac{x}{7} + 2 = -6$$

SOLUTION

Write the equation.
Subtract 2 from both sides of the equation.
Multiply each side by 7.

$$\frac{x}{7} + 2 = -6$$
$$\underline{\quad -2 \quad -2}$$
$$7 \cdot \frac{x}{7} \mid -8 \cdot 7$$
$$\boxed{x = -56}$$

✓ Check the solution.

$$\frac{-56}{7} + 2 \stackrel{?}{=} -6$$
$$-8 + 2 \stackrel{?}{=} -6$$
$$-6 = -6$$

$x = -56$

EXAMPLE 3

Solve the equation for x. Check the solution.
$$12 - 5x = 2$$

SOLUTION

Rewrite the equation by adding the opposite.
Subtract 12 from both sides.
Divide each side by (−5).

$$12 + (-5x) = 2$$
$$\underline{-12 \qquad -12}$$
$$\frac{-5x}{-5} = \frac{-10}{-5}$$
$$\boxed{x = 2}$$

✓ Check the solution.

$$12 - 5(2) \stackrel{?}{=} 2$$
$$12 - 10 \stackrel{?}{=} 2$$
$$2 = 2$$

$x = 2$

Lesson 4.3 ~ Solving Two-Step Equations

EXERCISES

Solve each equation for the variable. Show all work necessary to justify your answer.

1. $7x + 3 = 24$

2. $\frac{h}{2} + 1 = -4$

3. $\frac{x}{5} - 7 = 3$

4. $-3m + 6 = 30$

5. $6y + 1 = -17$

6. $0.25x + 10 = 12$

7. $14 = \frac{p}{4} + 10$

8. $\frac{1}{2}p - 3 = 9$

9. $\frac{w}{-4} + 8 = 3$

10. $\frac{a}{2} - 1.5 = 0.75$

11. $6 - 4w = 46$

12. $30x - 14 = 106$

13. $5 = -x - 4$

14. $-3 = 13 - 2d$

15. $\frac{f}{4} + \frac{3}{4} = 1\frac{1}{4}$

16. Juan solved three problems incorrectly. For each problem, describe the error he made and determine the correct answers.

a.
$$\begin{array}{r|r} -4x + 1 & = 21 \\ -1 & -1 \\ \hline \frac{-4x}{4} & = \frac{20}{4} \\ \\ x & = 5 \end{array}$$

b.
$$\begin{array}{r|r} 45 & = 20 - 5x \\ +20 & +20 \\ \hline \frac{65}{-5} & = \frac{-5x}{-5} \\ \\ x & = -13 \end{array}$$

c.
$$\begin{array}{r|r} \frac{x}{3} - 2 & = 22 \\ +2 & +2 \\ \hline 3 \div \frac{x}{3} & = 24 \div 3 \\ \\ x & = 8 \end{array}$$

Write an equation for each statement. Solve each equation. Show how you know each answer is correct.

17. Twice a number x, then increased by nine, is twenty-five. Find the number.

18. Four more than the quotient of the number y and negative three is eleven. Find the number.

19. Twelve decreased by the product of five and a number w is 52. Find the number.

20. Three less than one-half of a number p is 4. Find the number.

21. Jeannie got a new pair of eye glasses. She was able to pay $15 at the time she received her glasses. She will pay $12 each month on the balance. The original cost of the glasses was $87.
   a. Let x represent the number of months needed to pay off the cost of the glasses. Write an equation and explain how the equation represents this situation.
   b. How many months will it take Jeannie to pay for her glasses? Explain how you know your answer is correct.

22. Each month the fire department hosts a pancake feed. This month 75 people attended the pancake feed. This was 13 less than twice as many people as were at the pancake feed last month. How many people attended last month's pancake feed? Write and use an equation to find the answer.

23. Michael owns his own business selling hand-crafted birdhouses. He sells each birdhouse for $22. He charges $8 for shipping no matter how many birdhouses a person orders. A customer ordered birdhouses with a total bill of $206, including shipping. Write and solve an equation to determine how many birdhouses this customer ordered.

24. The perimeter of a rectangle is found using the formula $P = 2l + 2w$ where P is the perimeter, l is the length and w is the width. A rectangle has a perimeter of 70 centimeters and a side length of 12 centimeters. What is the width of the rectangle? Use numbers, words and/or an equation to show how you determined your answer.

REVIEW

Simplify each fraction.

25. $\frac{8}{10}$ **26.** $\frac{16}{32}$ **27.** $\frac{10}{75}$

28. $\frac{15}{60}$ **29.** $\frac{70}{120}$ **30.** $\frac{63}{84}$

Find the value of each integer expression.

31. $-2 - (-3)$ **32.** $-6(2)$ **33.** $11 + (-4)$

34. $-16 \div -8$ **35.** $-10(-3)$ **36.** $-4 + 4$

37. $9 - (-6)$ **38.** $-45 + (-21)$ **39.** $240 \div -20$

TIC-TAC-TOE ~ MATH DICTIONARY

Create a "Rational Numbers and Equations" dictionary. Locate all of the vocabulary words from all four Blocks in this textbook. Alphabetize the list of words and design a dictionary. The dictionary should include each word, spelled correctly, along with its definition. If appropriate, a diagram or illustration should be included.

THE DISTRIBUTIVE PROPERTY

LESSON 4.4

Use the Distributive Property to simplify expressions.
Solve equations using the Distributive Property.

very algebraic expression has at least one term. A **term** is a number or the product of a number and a variable. Terms are separated by addition and subtraction signs. A **constant** is a term that has no variable. The number multiplied by a variable in a term is called the **coefficient**.

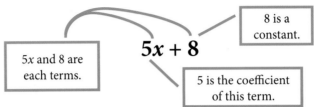

Some expressions contain parentheses. One tool that will help you work with these expressions is called the Distributive Property. The Distributive Property allows you to rewrite an expression without parentheses. This is done by distributing the front coefficient to each term inside the parentheses.

> **THE DISTRIBUTIVE PROPERTY**
>
> For any numbers a, b and c:
>
> $a(b + c) = a(b) + a(c)$ or $ab + ac$
> $a(b - c) = a(b) - a(c)$ or $ab - ac$

EXAMPLE 1   Use the Distributive Property to simplify each expression.
a. $7(x + 8)$ b. $-3(y - 2)$ c. $\frac{1}{2}(4x - 3)$

SOLUTIONS   Drawing arrows from the front coefficient to each term inside the parentheses will help guide you.

a. $7(x + 8) = 7(x) + 7(8)$
$= 7x + 56$

b. $-3(y - 2) = -3(y) - (-3)(2)$
$= -3(y) - (-6)$
$= -3y + 6$

c. $\frac{1}{2}(4x - 3) = \frac{1}{2}(4x) - \frac{1}{2}(3)$
$= 2x - 1\frac{1}{2}$

Multiply the two coefficients together. Think of it as $\frac{1}{2} \cdot 4 \cdot x$.

Equations that contain parentheses can be solved by following a process that includes using the Distributive Property and inverse operations.

SOLVING EQUATIONS USING THE DISTRIBUTIVE PROPERTY

1. Distribute the front coefficient to remove the parentheses.
2. Undo addition or subtraction using inverse operations.
3. Undo multiplication or division using inverse operations.
4. Check your answer.

EXAMPLE 2

Solve each equation for the variable. Check your solution.
a. $3(x + 5) = 33$
b. $-5(2m - 1) = 25$

SOLUTIONS

a. Write the equation. $3(x + 5) = 33$
Distribute 3 through the parentheses. $3x + 15 = 33$
Subtract 15 from each side of the equation. $-15 \quad -15$
Divide both sides by 3. $\dfrac{3x}{3} = \dfrac{18}{3}$
$x = 6$

☑ Check the solution.
$3(6 + 5) \stackrel{?}{=} 33$
$3(11) \stackrel{?}{=} 33$
$33 = 33$

b. Write the equation. $-5(2m - 1) = 25$
Distribute (-5) through the parentheses. $-10m + 5 = 25$
Subtract 5 from each side of the equation. $-5 \quad -5$
Divide both sides by -10. $\dfrac{-10m}{-10} = \dfrac{20}{-10}$
$m = -2$

☑ Check the solution.
$-5(2(-2) - 1) \stackrel{?}{=} 25$
$-5(-4 - 1) \stackrel{?}{=} 25$
$-5(-5) \stackrel{?}{=} 25$
$25 = 25$

Remember the Distributive Property by thinking of careers where people distribute things.

EXAMPLE 3

Tiffany works as a travel agent. All the employees at her work are required to sell a certain number of vacation packages each day. Tiffany has sold 3 more than the required amount each day for the last 12 days. If she has sold 84 vacation packages, what is the required amount of packages that each employee must sell daily?

SOLUTION

Let v represent the required number of vacation packages sold daily.

Write the equation. $12(v + 3) = 84$
Distribute 12 through the parentheses. $12v + 36 = 84$
Subtract 36 from both sides. $-36\quad -36$
Divide both sides by 12. $\dfrac{12v}{12} = \dfrac{48}{12}$
$v = 4$

☑ Check the solution.
$12(4 + 3) \stackrel{?}{=} 84$
$12(7) \stackrel{?}{=} 84$
$84 = 84$

Each employee must sell 4 vacation packages daily.

EXERCISES

Use the Distributive Property to rewrite each expression without parentheses.

1. $8(x + 5)$ **2.** $3(y - 2)$ **3.** $-6(m - 4)$

4. $7(2x - 3)$ **5.** $-2(5 + 6p)$ **6.** $\frac{1}{3}(9x - 6)$

7. $0.4(2y + 9)$ **8.** $-3(10 - x)$ **9.** $11(2m - 4)$

10. Chloe used the Distributive Property to rewrite the expression $7(x - 3)$ as $7x - 3$. Explain her error in a complete sentence. Find the correct expression.

11. Three friends went out to lunch. Each person bought a sandwich and juice. The sandwiches were $3.50 each and each bottle of juice cost $1.25. Show how you could use the Distributive Property to find the total cost for the three lunches.

Solve each equation. Show all work necessary to justify your answer.

12. $3(x + 4) = 18$

13. $-5(x + 3) = 20$

14. $24 = 2(x - 7)$

15. $-7(2x - 4) = -42$

16. $\frac{1}{4}(4x + 10) = 8$

17. $4 = -4(x - 1)$

18. $3(3x - 7) = -30$

19. $11 = -4(x + 1)$

20. $0.5(8x + 2) = 17$

21. One year the Northwest suffered from a drought. The total snow pack at the peak of the Cascades was only 30 inches in February. For the 60 days prior to this measurement, the average snowfall was 2.5 inches less than the normal average per day.
  a. The equation $60(x - 2.5) = 30$ represents this situation. What does x represent in the equation?
  b. Solve the equation and determine what the normal average daily snowfall is at the peak of the Cascades.

22. Three friends went to the roller skating rink together on Friday night. It cost $4.30 for each person to enter the rink. Each friend spent an equal amount of money at the concession stand. The evening cost the three friends a total of $19.50.

  a. Let c represent the amount of money each friend spent at the concession stand. Write an equation to represent this situation.
  b. How much did each friend spend at the concession stand? Show all work necessary to justify your answer.

REVIEW

Find the value of each rational expression.

23. $\frac{3}{4} + \left(-\frac{1}{2}\right)$

24. $1\frac{2}{5} - \left(-\frac{1}{10}\right)$

25. $-\frac{5}{6}\left(-\frac{3}{10}\right)$

26. $3\frac{1}{3} \div \frac{1}{5}$

27. $-\frac{5}{8} + \left(-\frac{1}{4}\right)$

28. $\left(3\frac{1}{6}\right)\left(2\frac{2}{5}\right)$

TIC-TAC-TOE ~ LETTER TO THE EDITOR

Each state has high school diploma requirements in mathematics. Research the diploma requirements in mathematics for your state. How many math credits are needed? Is there a restriction on the level of math? Do you have to pass an exit test?

Write a letter to the editor of your town explaining why you believe the graduation requirements will help students like you and your friends after you graduate from high school. Use real-world examples and situations to support your reasons.

SIMPLIFYING EXPRESSIONS

LESSON 4.5

 Simplify expressions using the Distributive Property and combining like terms.

Like terms are terms that contain the same variable(s). The coefficients do not need to be the same in order for the terms to be like terms. Like terms can be combined by adding or subtracting the coefficients.

LIKE TERMS

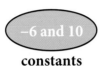
constants

terms with x

terms with y

To **simplify an expression**, write it so it has no parentheses and all like terms have been combined. When combining like terms, you must remember that the operation in front of the term (addition or subtraction) must stay attached to the term. Rewrite the expression by grouping like terms together before adding or subtracting the coefficients to simplify.

$$-3x + 5x = (-3 + 5)x = 2x$$

EXAMPLE 1 Simplify each algebraic expression by combining like terms.
a. $4x - 7x + 5x$
b. $8y + (-6) + 3 - 4y$

SOLUTIONS

a. All terms are alike. Add and subtract the coefficients from left to right.

$4x - 7x + 5x = (4 - 7 + 5)x$
$= 2x$

b. There are two types of like terms in this algebraic expression. There are constants and terms that have the y variable.

Rewrite the expression so the like terms are next to each other.

$8y + (-6) + 3 - 4y$
$8y - 4y + (-6) + 3$

Add the like terms together. $4y$ -3

$8y + (-6) + 3 - 4y = 4y - 3$

Remember to move the operation with the term: $-4y$

Can also be written as $4y + (-3)$

126 Lesson 4.5 ~ Simplifying Expressions

EXAMPLE 2 Simplify $9m + 3p - 3p - 2m + m$.

SOLUTION

Group the like terms together. The addition or subtraction sign must stay attached to the term.

Add or subtract the coefficients of the like terms from left to right.

$0p$ represents zero p's so the term does not need to be written.

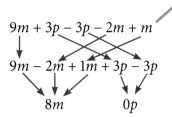

The coefficient of m is 1.

$9m + 3p - 3p - 2m + m$
$9m - 2m + 1m + 3p - 3p$
$\quad\quad 8m \quad\quad\quad 0p$

$8m + 0p = 8m$

$9m + 3p - 3p - 2m + m = \boxed{8m}$

SIMPLIFYING ALGEBRAIC EXPRESSIONS

1. Rewrite the expression without parentheses using the Distributive Property, if needed.
2. Combine all like terms.

EXAMPLE 3 Simplify each algebraic expression.
  a. $4(x + 5) - 3x + 7$ **b.** $6y - 2(y - 7) + 3$

SOLUTIONS

a. Distribute 4 through the parentheses.

$\quad$ Group like terms.
$\quad$ Combine like terms.

$4(x + 5) - 3x + 7$
$4x + 20 - 3x + 7$
$4x - 3x + 20 + 7$
$x + 27$

b. Distribute (-2) through the parentheses.

$\quad$ Combine like terms.

$6y - 2(y - 7) + 3$
$6y - 2y + 14 + 3$
$4y + 17$

In each example so far in this lesson, the original and simplified expressions are called **equivalent expressions**. Equivalent expressions are expressions which represent the same algebraic expression.

EXPLORE! WHERE DO I BELONG?

Step 1: Copy each expression below on your own paper, leaving two lines between each expression.

- A. $6(x - 4) - 2x + 14$
- B. $3(2x - 3) - 3x$
- C. $9 - 2(x + 1)$
- D. $-5x + 4(x + 3) - 11$
- E. $5x + 3 - x - 13$
- F. $2x + 5 + (-4) - 3x$
- G. $1 + 2(5x - 5) - 7x$
- H. $-3x + 8 + x - 1$
- I. $-2(x - 3) + x - 5$
- J. $7(x + 2) - 9(x + 1) + 2$
- K. $5x + 2x - x - 3x - 9$
- L. $2(4x - 5) - 4x$

Step 2: Simplify each expression.

Step 3: Every expression listed above is equivalent to two other expressions in the list. Classify the twelve expressions into four groups of equivalent expressions.

Step 4: Create another "non-simplified" expression for each group that is equivalent to the other expressions in the group. Show how you know your expression fits with the others in the group.

EXERCISES

1. Write an expression that meets each condition:
 a. has three terms, none of which are like terms.
 b. has four terms that are like terms.

2. Describe in words how you would simplify $5(x + 3) + 2x$.

3. Examine the expression $6 + 5y + x - 8y$.
 a. Which terms are like terms?
 b. What is the coefficient of the x term?
 c. Write the expression in simplified form.

Simplify each algebraic expression.

4. $5y - 9y + 6x - 2y$

5. $x + y + x - 2y - x$

6. $11 + 5d - d - 4 - 7$

7. $6m + 30 - 2m + 11 - m$

8. $4w + 1 + w + 3w - 7$

9. $2(x + 9) - 15$

10. $8(2x - 1) - 10x + 8$

11. $4(p + 3) + 5$

12. $7 - 3(x - 2)$

13. $5 + 3(2x + 4)$

14. $2(x + 1) + 4(x - 1)$

15. $10y - 2y + 3(y - 7)$

16. $-5(2f - 7) + 10f - 30$

17. $\frac{1}{2}(6x + 3) + \frac{1}{2}$

18. $4(0.4y - 2.1) + 3.6 - y$

19. It takes Gregory 50 minutes to get to his best friend's house. He spends m minutes riding his bike to the bus stop at a rate of 0.25 miles per minute. He gets on the bus and travels the rest of the way at a rate of 0.6 miles per minute. The total distance traveled is represented by the expression $0.25m + 0.6(50 - m)$.
    a. Explain what $(50 - m)$ represents in this situation.
    b. Simplify this expression.
    c. Greg rides his bike for 10 minutes. How many miles did he ride? Show all work necessary to justify your answer.

20. Jennifer and Julius disagree about whether or not the expression $2x^2 + 4x + 7$ can be simplified. Jennifer thinks the first two terms can be combined but Julius says they are not alike because one x-variable is squared and the other is not. Who do you agree with and why?

21. Write an expression with at least six terms that simplifies to $-3x + y$. Show all work necessary to justify your answer.

REVIEW

Solve each equation for the variable. Show all work necessary to justify your answer.

22. $\frac{1}{5}x = -7$ **23.** $10.7 = -3.5 + d$ **24.** $\frac{b}{4} = -0.24$

25. $-3x + 8 = 35$ **26.** $-4 + \frac{y}{2} = -1$ **27.** $8(x - 2) = -40$

TIC-TAC-TOE ~ NATE'S BUSINESS

Nate started a business selling coffee and muffins at the local senior center. He spent $250 for a coffee cart and also paid $500 to rent a space for his cart at the senior center for one year. He calculated that he makes a profit of $0.75 for each cup of coffee and $1.25 for each muffin he sells. Help evaluate Nate's business plan. Show all work.

   1. If Nate were to only sell coffee and no muffins, how many cups of coffee would he need to sell in one year to break even (cover his expenses)?

2. If Nate were to sell the same amount of coffee and muffins, about how many of each would he need to sell to break even?

3. Nate estimates he will make a profit of about $945 this year.
    a. If he only sells coffee, how many cups will he need to sell to reach his estimate?
    b. If he only sells muffins, how many muffins will he need to sell to reach his estimate?
    c. If he sells both muffins and coffee, determine two different combinations of coffee and muffins that would allow him to reach a profit of $945.

4. Design your own business. What would your expected start-up costs be? What products would you sell? How much would you hope your profit would be for each product you would sell? How many of each product would you need to sell to break even?

SIMPLIFYING AND SOLVING EQUATIONS

LESSON 4.6

  Simplify and solve equations with variables on one side of an equation.

Some equations must be simplified before you can perform inverse operations to isolate the variable. Sometimes equations have like terms on one side that must be combined first before solving.

EXPLORE!   **EQUATION MANIPULATION**

Step 1: If you do not have an equation mat, draw one on a blank sheet of paper. On your equation mat, model the equation $4x + 7 + x - 3 = -1$.

Step 2: Looking at your equation mat, is there a way you can simplify either side of the equation? If so, rewrite the equation.

Step 3: Solve the equation using your mat. What does x equal?

Step 4: Look at the equation mat at the right and determine which of the equations below is represented on the mat. Explain why you chose the one you did and find the solution to the equation.

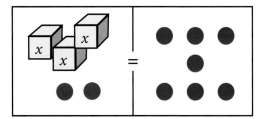

    Equation #1: $-2 + 2x + 4 + 2x = 7$
    Equation #2: $-7 + 2x + 5 + x = 7$
    Equation #3: $3x + 2x - 3 + 1 = 7$

Step 5: Explain in words how to simplify an algebraic equation before solving.

Step 6: Use the equation mat to solve the equation shown on the mat above.

SIMPLIFYING AND SOLVING EQUATIONS WITH LIKE TERMS

1. Using the Distributive Property, rewrite the equation without parentheses.
2. Combine all like terms that are on the same side of the equation.
3. Solve the remaining one- or two-step equation by using inverse operations to isolate the variable.

Simplify each side of an equation before solving!

EXAMPLE 1 Solve each equation for the variable. Check your solution.
a. $3m + 6 - 8m + 9 = -20$
b. $0 = 4(3x - 2) - 5x + x$

SOLUTIONS

a. Write the equation. $3m + 6 - 8m + 9 = -20$
Group like terms. $3m - 8m + 6 + 9 = -20$
Combine like terms. $-5m + 15 = -20$
Subtract 15 from both sides. $\underline{-15\ \ -15}$
Divide both sides of the $\dfrac{-5m}{-5} = \dfrac{-35}{-5}$
equation by (−5).
$$m = 7$$

☑ Check the solution.
$3(7) + 6 - 8(7) + 9 \stackrel{?}{=} -20$
$21 + 6 - 56 + 9 \stackrel{?}{=} -20$
$-20 = -20$

b. Write the equation. $0 = 4(3x - 2) - 5x + x$
Distribute 4 through the parentheses. $0 = 12x - 8 - 5x + x$
Group like terms. $0 = 12x - 5x + x - 8$
Combine like terms. $0 = 8x - 8$
Add 8 to both sides of the equation. $\underline{+8 +8}$
Divide both sides by 8. $\dfrac{8}{8} = \dfrac{8x}{8}$
$$1 = x$$

☑ Check the solution.
$0 \stackrel{?}{=} 4(3 \cdot 1 - 2) - 5(1) + 1$
$0 \stackrel{?}{=} 4(1) - 5 + 1$
$0 = 0$

EXAMPLE 2 Natasha paid $60 to join a gymnastics club for the summer. Each gymnastics class she attended costs $15. Her aunt works at the club so Natasha receives a $4 "Friends and Family" discount per class. Natasha spent $214 on classes for the entire summer. How many classes did she attend?

SOLUTION Let g represent the number of classes which Natasha attended. An equation which models this situation could be written as:

$$60 + 15g - 4g = 214$$

Combine like terms. $60 + 11g = 214$
Subtract 60 from both sides. $\underline{-60 -60}$
Divide each side by 11. $\dfrac{11g}{11} = \dfrac{154}{11}$
$$g = 14$$

Natasha attended 14 gymnastics classes.

Lesson 4.6 ~ Simplifying and Solving Equations

EXERCISES

1. In words, describe the steps you would take to solve the equations.
   a. $4x - 2 + 6x + 15 = 38$
   b. $5(x + 2) + 3x = 82$.

Solve each equation for the variable. Show all work necessary to justify your answer.

2. $8x - 3x + 4x = -36$

3. $6y - 4 - 3y + 1 = 12$

4. $-21 = 2(x - 8) + 3x$

5. $10 + 15(w - 2) = 40$

6. $k + 2k + 3k - k = -100$

7. $0.5(x + 4) + 0.25x = 8$

8. $0.2y + 1.4y - 0.5y + 1.4 = 8$

9. $73 = 3(2x + 5) - 2(x - 5)$

10. $-2(d + 20) - 8d = 20$

11. $\frac{1}{4}(w + 4) + \frac{1}{2}w = 10$

12. $0 = 13 - 2x + 5x - 11$

13. $5 - 3p + 2p = -10$

14. The Mount Joseph Fall Festival is held each year as a fundraiser for children with cancer. Each booth at the festival charges a set price to play. Timothy took $32 to the Festival. He played games at 6 booths, took a break, and then played games at 5 more booths. He had $7.25 remaining at the end of the evening.
    a. Let b represent the cost to play a game at each booth. Explain how the equation $32 - 6b - 5b = 7.25$ models this situation.
    b. Solve the equation. How much did it cost to play a game at one booth at the Festival? Explain how you know your answer is correct.

15. Mykisha told her friend this riddle: "I am thinking of a number. Half of my number plus twice my number is twenty. What number am I thinking of?"
    a. Write an equation that models Mykisha's statement.
    b. Solve the equation. What number was Mykisha thinking of?

16. Larry's math teacher did not want to reveal her age. She told her students if they could solve her riddle, it would reveal how old she is. Her riddle stated: "The product of my age and five, then decreased by twenty and finally increased by my age times the opposite of three is fifty." How old is Larry's math teacher? Use mathematics to justify your answer.

17. James went to Central Park one Saturday in October with his cross-country team for a 90 minute workout. He ran m minutes at a rate of 0.15 miles per minute. During the time he was not running, he walked at a rate of 0.06 miles per minute. He totaled 9.9 miles. This situation is represented by the equation: $0.15m + 0.06(90 - m) = 9.9$.
    a. What does $(90 - m)$ represent in this situation?
    b. Solve the equation to determine how many minutes James ran. Explain how you know your answer is correct.

REVIEW

Find the value of each expression.

18. $9 + \frac{17-13}{3-1}$

19. $-7(2-5) - 4$

20. $1 + (5+1)^2$

21. $\frac{(1-4)^2}{2+1} - 2(-5)$

22. $-20(3) - 10(-4)$

23. $\frac{-50}{(-7+2)^2} + 12$

Insert one set of parentheses in each numerical expression so that it equals the stated amount. Use mathematics to show that your answer is correct.

24. $3 \cdot 7 + 2 + 1 \cdot 5 = 36$

25. $-4 + 1 \cdot 3 - 2 \cdot 3 = -15$

26. $5 + -2 \cdot 6 + 1 = 19$

TIC-TAC-TOE ~ ABSOLUTE VALUE

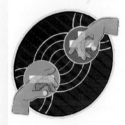

Absolute value is the distance a value is from 0 on a number line. Since the distance from 0 is represented by a positive number, the absolute value of an expression is always positive. When solving absolute value equations, there are often two answers that make the equation true. Take the expression inside the absolute value bars and set it equal to the given amount. Also, set it equal to the opposite of the given amount.

For example: Solve for x: $|2x + 7| = 13$

Write two equations to solve:

$2x + 7 = 13$ and $2x + 7 = -13$

$x = 3$ and $x = -10$

✓ $|2(3) + 7| = 13$ and $|2(-10) + 7| = 13$

Solve each absolute value equation for both solutions. Show all work necessary to justify your answer.

1. $|3x| = 21$

2. $\left|\frac{x}{7}\right| = 1$

3. $\left|\frac{x}{2-6}\right| = 3$

4. $25 = |-4x + 1|$

5. $\left|\frac{1}{2}x - 5\right| = 7$

6. $|2x + 3x + x| = 48$

7. $\left|-10 + \frac{x}{3}\right| = 2$

8. $36 = |4(x - 7)|$

9. $|-0.25x + 2| = 5$

Lesson 4.6 ~ Simplifying and Solving Equations

SOLVING EQUATIONS WITH VARIABLES ON BOTH SIDES

LESSON 4.7

Simplify and solve equations with variables on both sides of an equation.

Shannon and Matt each started a savings account on the same day. Shannon opened her account with $20 and puts an additional $12 in her account each week. Matt opened his account with $55 and deposits $5 in his account each week. On what week will Matt and Shannon have the same amount of money in their accounts?

Let w represent the number of weeks that have passed. The expression that represents Shannon's savings is $20 + 12w$. The expression that represents Matt's savings is $55 + 5w$. Setting the two equations equal to each other represents the time when both accounts will have the same amount of money.

$$20 + 12w = 55 + 5w$$

To solve this equation, you must first get the variables on the same side of the equation. To do this, move one variable term to the opposite side of the equation using inverse operations. It is easiest to move the variable term that has the smaller coefficient so that you work with fewer negative numbers. After the variables are on the same side of the equation, you must solve the two step equation that remains.

$$
\begin{array}{r}
20 + 12w = 55 + 5w \\
\underline{-5w -5w} \\
20 + 7w = 55 \\
\underline{-20 -20} \\
\dfrac{7w}{7} = \dfrac{35}{7} \\
w = 5
\end{array}
$$

5 is a smaller coefficient than 12 so $5w$ will be moved.

Shannon and Matt will have the same amount in their savings accounts after 5 weeks.

To check the solution, substitute the value into each side of the equation. If both sides are equal, the solution is correct.

Shannon's Savings Account after 5 weeks: $20 + 12(5) = \$80$
Matt's Savings Account after 5 weeks: $55 + 5(5) = \$80$

SOLVING EQUATIONS WITH VARIABLES ON BOTH SIDES

1. Move one variable term to the opposite side of the equals sign using inverse operations.
2. Move all constants away from the variable using inverse operations.
3. Use division to isolate the variable.

EXAMPLE 1

Solve each equation for the variable. Check the solution.
a. $3x - 26 = 7x + 2$
b. $6m + 22 = -2m + 10$

SOLUTIONS

a. Write the equation.
Subtract $3x$ from both sides.

$$3x - 26 = 7x + 2$$
$$\underline{-3x \qquad\quad -3x}$$
$$-26 = 4x + 2$$

Subtract 2 from each side of the equation.
Divide both sides by 4.

$$\underline{-2 \qquad -2}$$
$$\frac{-28}{4} = \frac{4x}{4}$$
$$\boxed{-7 = x}$$

☑ Check the solution

$$3(-7) - 26 \stackrel{?}{=} 7(-7) + 2$$
$$-21 - 26 \stackrel{?}{=} -49 + 2$$
$$-47 = -47$$

$-2m$ is the same as subtracting $2m$, so the inverse operation is adding $2m$.

b. Write the equation.
Add $2m$ to both sides.

$$6m + 22 = -2m + 10$$
$$\underline{+2m \qquad\quad +2m}$$
$$8m + 22 = 10$$

Subtract 22 from each side of the equation.
Divide both sides by 8.

$$\underline{-22 \quad -22}$$
$$\frac{8m}{8} = \frac{-12}{8}$$
$$\boxed{m = -1.5}$$

☑ Check the solution.

$$6(-1.5) + 22 \stackrel{?}{=} -2(-1.5) + 10$$
$$-9 + 22 \stackrel{?}{=} 3 + 10$$
$$13 = 13$$

EXAMPLE 2

The Underwood family is 340 miles away from home and are headed toward home at a rate of 52 miles per hour. The Underwoods' next-door neighbors, the Jacksons, are leaving home, traveling toward the Underwoods at a rate of 48 miles per hour. How long will it be before they pass each other?

SOLUTION

Let h represent the number of hours the two families have been on the road. The expression that represents the Underwoods' distance from home is $340 - 52h$. The expression that represents the Jacksons' distance from home is $48h$.

Set the two expressions equal to one another.
Add $52h$ to both sides.
Divide each side by 100.

$$340 - 52h = 48h$$
$$\underline{+52h \quad +52h}$$
$$\frac{340}{100} = \frac{100h}{100}$$
$$\boxed{3.4 = h}$$

The Jackson and Underwood families will pass each other after 3.4 hours.

EXERCISES

Solve each equation for the variable. Show all work necessary to justify your answer.

1. $5x + 6 = 7x$

2. $2p + 10 = 8p - 14$

3. $12 - 3c = c$

4. $-1.2 + 0.8m = m - 1$

5. $y + 4 = 3y + 3$

6. $-8w + 30 = 2w - 20$

7. $2h = 15 - 3h$

8. $x + 6 = 2x + 2$

9. $12.4 + 2a = 5a + 5.5$

10. $-3p + 5 = -7p - 11$

11. $\frac{2}{3}w + 1 = \frac{5}{6}w + 1\frac{1}{2}$

12. $14x - 14 = 2x + 46$

13. $7 - x = -5x - 1$

14. $-11 + 2d = 13 - 2d$

15. $\frac{1}{2}y + 2 = \frac{1}{4}y + 5$

16. In words, describe the steps you would take to solve the equation $8 + 5x = -6x + 30$.

Write an equation for each statement. Solve each problem. Show all work necessary to justify your answer.

17. Three times a number x plus ten is five times the number. Find the number.

18. The number y times negative four plus five is the number. Find the number.

19. Twelve decreased by twice a number w is equal to eight less than three times w. Find the number.

20. JJ's Fitness Club offers two different fees for their yoga classes. Club members are charged a one-time membership fee of $24 and pay $3 per class. Non-members pay $7 per class. Let y represent the number of yoga classes attended.
    a. Write an expression to represent the cost for a member to attend y classes.
    b. Write an expression to represent the cost for a non-member to attend y classes.
    c. Set the two expressions equal to each other and solve the equation to determine how many classes result in the same cost for a member and non-member.

21. Christopher is at a friend's house 4 miles away from his home. At 5:00 PM, he begins riding home on his bicycle at a rate of 0.3 miles per minute. At the same time, his mom decides to go pick him up. She leaves their home in her car traveling at 0.5 miles per minute. Let m represent the number of minutes both have been on the road.
    a. Write two different expressions, one to represent Christopher's distance from home and the other to represent his mom's distance from home.
    b. How many minutes will it take before Christopher and his mom cross paths? Explain how you know your answer is correct.

22. Sarah begins the year with $100 in her savings account. Each week, she spends $8. Martin begins the year with no money saved, but each week he puts $12 in an account. Let x represent the number of weeks that have passed since the beginning of the year. How many weeks will have passed when they have the same amount of money in their accounts? Show all work necessary to justify your answer.

23. The local Store 'n' More has a coffee club. If you join the club for a one-time fee of $12, it only costs $3 to buy a latte. If you choose not to join the club, it costs $4.50 to purchase a latte. After how many lattes will the two plans cost the same amount? Use words and/or numbers to show how you determined your answer.

REVIEW

Solve each equation for the variable. Show all work necessary to justify your answer.

24. $4x + 11 = -33$

25. $6(y + 1) = 36$

26. $26 = 5(x + 10) + 3x$

27. $\frac{y}{7} - 6 = 2$

28. $5m + 2m + m = 80$

29. $12 = \frac{2}{3}(6x + 12)$

30. $8 - 2(w - 6) = 30$

31. $7 - k + 3 = 10$

32. $0.3(x + 5) + 0.5 = 8$

Find each sum, difference, product or quotient.

33. $\frac{9}{10} + \frac{1}{3}$

34. $\frac{3}{4} \div \frac{15}{28}$

35. $2\frac{2}{3} - 1\frac{3}{4}$

36. $6\left(\frac{5}{12}\right)$

37. $-\frac{5}{6} + \frac{1}{5}$

38. $-5\frac{1}{2}\left(1\frac{1}{4}\right)$

TIC-TAC-TOE ~ EQUATION POSTER

To solve a multi-step equation you must follow a step-by-step process. Design a poster that could be displayed in your classroom that explains how to solve multi-step equations with variables on both sides of the equals sign. Include both written instructions and examples. Make a list of what you think the most common errors might be on the poster so students will know what mistakes to watch for.

TIC-TAC-TOE ~ EQUATION MATS

Equation mats are used to see a visual model of the equation-solving process. Write a "How To..." guide on using equation mats for a variety of different types of equations. Include the following types of equations in your guide:
- One-Step Equations
- Two-Step Equations
- Equations with Variables on Both Sides of the Equals Sign

LINEAR INEQUALITIES

LESSON 4.8

 Solve inequalities with one variable.

Jennifer has less than $75 in her bank account. Monte has at least 300 baseball cards in his collection. Each of these statements can be written using an **inequality**. Inequalities are mathematical statements which use $>$, $<$, $\geq$ or $\leq$ to show a relationship between quantities.

Jennifer has less than $75 in her bank account.

$$j < \$75$$

Monte has at least 300 baseball cards.

$$m \geq 300$$

Inequalities have multiple answers that can make the statement true. In Jennifer's example, she might have $2 or $74. All that is known for certain is that she has under $75 in her account. In Monte's example, he might have 300 cards (this comes from the "equal to" part of $\geq$) or 3,000,000 cards. There is an infinite number of possibilities that would make this statement true.

INEQUALITY SYMBOLS

$>$	"greater than"
$<$	"less than"
$\geq$	"greater than or equal to"
$\leq$	"less than or equal to"

EXAMPLE 1 Write an inequality for each statement.
a. Nanette's height (h) is greater than 50 inches.
b. Pete has at least $6 in his pocket. Let p represent the amount of money.
c. Jeremy is less than 20 years old. Let j represent Jeremy's age.

SOLUTIONS

a. The key words are "greater than". Use the symbol $>$. $h > 50$

b. The key words are "at least". This means the lowest amount he could have is $6. Use the $\geq$ symbol. $p \geq 6$

c. The key words are "less than". Use the $<$ symbol. $j < 20$

Solutions to an inequality can be graphed on a number line. When using the > or < inequality symbols, an "open circle" is used on the number line because the solution does not include the given number. For example, if $x > 2$, the solution cannot include 2 because 2 is not greater than 2. When using the $\geq$ or $\leq$ inequality symbols, a "closed (or filled in) circle" is used because the solution contains the given number.

Determining which direction the arrow should point is based on the relationship between the variable and the solution. When graphing an inequality on a number line, the solution is shown by using an arrow to point towards the set of numbers that make the statement true. Notice in the box below how the circle is either an open or closed circle to designate the type of inequality that is being graphed.

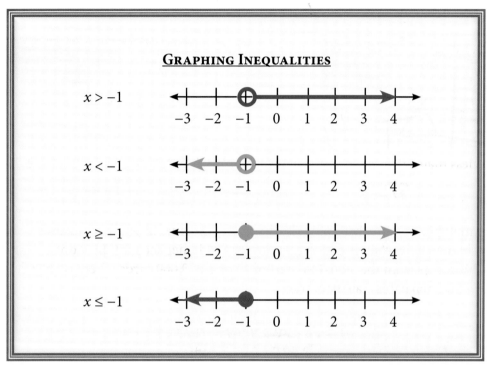

Inequalities are solved using properties similar to those you used to solve equations. Use inverse operations to isolate the variable so the solution can be graphed on a number line.

EXAMPLE 2 Solve the inequality and graph its solution on a number line.
$$3x - 1 < 14$$

SOLUTION

Write the inequality.
Add 1 to both sides of the inequality.
Divide both sides of the inequality by 3.

$$3x - 1 < 14$$
$$ +1 \phantom{<} +1$$
$$\frac{3x}{3} < \frac{15}{3}$$
$$x < 5$$

Graph the solution on a number line. Use an open circle.

Lesson 4.8 ~ Linear Inequalities

EXAMPLE 3   Solve the inequality and graph its solution on a number line.
$$5x + 1 \leq 3x - 7$$

SOLUTION   Write the inequality.
Subtract $3x$ from each side of the inequality.

Subtract 1 from each side.

Divide both sides by 2.

$$\begin{aligned} 5x + 1 &\leq 3x - 7 \\ -3x & -3x \\ \hline 2x + 1 &\leq -7 \\ -1 & -1 \\ \hline \frac{2x}{2} &\leq \frac{-8}{2} \\ x &\leq -4 \end{aligned}$$

Graph the solution on a number line. Use a closed circle.

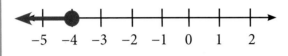

One special rule applies to solving inequalities. Whenever you multiply or divide by a negative number on both sides of the equation, you must flip the inequality symbol. For example, less than (<) would become greater than (>) if you multiply or divide by a negative when performing inverse operations.

EXAMPLE 4   Solve the inequality $-4x + 7 \leq 19$.

SOLUTION   Write the inequality.
Subtract 7 from each side of the inequality.
Divide both sides by -4.

Since both sides were divided by a negative, flip the inequality symbol.

$$\begin{aligned} -4x + 7 &\leq 19 \\ -7 & -7 \\ \hline \frac{-4x}{-4} &\leq \frac{12}{-4} \\ x &\geq -3 \end{aligned}$$

The sign changed direction because both sides were divided by a negative number.

EXERCISES

Write an inequality for each graph. Use x as the variable.

1.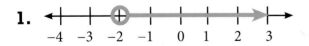

2.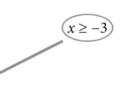

3.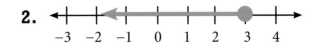

4.

Solve each inequality and graph each on a number line.

5. $5x \geq -20$

6. $x + 7 < 6$

7. $10 < 3x + 1$

8. $2x + 7 > 15$

9. $\frac{x}{2} - 1 \geq -4$

10. $-3x - 4 < 5$

11. $5(x + 3) \leq 20$

12. $7 > \frac{x}{-4} + 6$

13. $9x < 2x - 35$

14. $-7 + 4x \geq 3 - 6x$

15. $2(x + 3) \geq 5x + 12$

16. $\frac{1}{2}x + 8 < x + 4$

17. A cargo elevator has a maximum carrying capacity of 240 pounds. Each cargo box weighs 20 pounds.
  a. Write and solve an inequality that represents the maximum number of cargo boxes that the elevator can hold.
  b. A 140-pound person rides in the elevator with the cargo boxes. What is the maximum number of cargo boxes the person can take with them on the elevator? Show that your answer is correct by showing that one more than your answer would exceed the elevator capacity.

18. Frankie has $400 in her bank account at the beginning of the summer. She wants to have at least $150 in her account at the end of the summer. Each week she withdraws $22 for food and entertainment.
  a. Write an inequality for this situation. Let *x* represent the number of weeks she withdraws money from her account.
  b. What is the maximum number of weeks that Frankie can withdraw money from her account? Explain how you know your answer is correct.

19. Ethan was at the beach. He wanted to spend $10 or less on a beach bike rental. The company he chose to rent from charged an initial fee of $4 and an additional $0.35 per mile he rode.
  a. Write an inequality for this situation. Let *x* represent the number of miles ridden.
  b. How many miles can Ethan ride without going over his spending limit? Round to the nearest whole mile.

20. Callie and BreShay went to the mall. They each spent the exact same amount during the day. Callie spent less than or equal to $45. BreShay spent more than $40. Create a number line that shows all the possible amounts that Callie or BreShay could have spent during the day. Use words and/or numbers to show how you determined your answer.

REVIEW

Solve each equation. Show all work necessary to justify your answer.

21. $2x + 5 = 17$

22. $3y - 1 + 2y = 49$

23. $5 + \frac{m}{10} = -2$

24. $10(x + 1) = -27$

25. $2(x + 8) = 8(x + 1)$

26. $\frac{1}{2}x - 9 = \frac{3}{4}x - 17$

REVIEW

BLOCK 4

Vocabulary

algebraic expression	equivalent expressions	simplify an expression
coefficient	evaluate	solution
constant	inequality	term
equation	like terms	variable

Write and evaluate algebraic expressions.
Determine if a number is a solution of an equation.
Use inverse operations to solve one-step equations.
Solve two-step equations.
Use the Distributive Property to simplify expressions.
Solve equations using the Distributive Property.
Simplify expressions using the Distributive Property and combining like terms.
Simplify and solve equations with variables on one side of an equation.
Simplify and solve equations with variables on both sides of an equation.
Solve inequalities with one variable.

Lesson 4.1 ~ Expressions and Equations

Write an algebraic expression for each phrase.

1. the product of nine and y

2. the sum of m and six

3. four less than the product of z and eight

4. two more than the quotient of c and five

Evaluate each expression.

5. $\frac{x}{2} - 9$ when $x = 30$

6. $6b + 5$ when $b = -4$

7. $5k - 7n$ when $k = 10$ and $n = -2$

8. $\frac{2}{9}m + \frac{1}{3}$ when $m = 3$

Determine if the number given is the solution to the equation.

9. $4x + 1 = 9$ Is 2 the solution?

10. $-3y - 5 = -7$ Is 4 the solution?

11. $\frac{m}{6} + 5 = 7.3$ Is 15 the solution?

12. $\frac{1}{4}h + 2 = 0$ Is −8 the solution?

Lesson 4.2 ~ Solving One-Step Equations

Solve each equation using an equation mat or inverse operations. Show all work necessary to justify your answer.

13. $3x = 21$

14. $y + 81 = 107$

15. $\dfrac{p}{-4} = 7$

16. $3.25 = m - 6.25$

17. $0.2k = 1.1$

18. $-\dfrac{5}{12} + x = \dfrac{1}{6}$

Write an algebraic equation for each sentence. Solve the equation using an equation mat or inverse operations.

19. The product of x and nine equals thirty-six.

20. The quotient of x and two is two-fifths.

21. Thirty-one less than x is negative ten.

22. The sum of x and one-half equals thirteen.

23. Kate is thinking of two numbers. The sum of the numbers is −8. If one of the numbers is 12, what is the other number? Use words and/or numbers to show how you determined your answer.

Lesson 4.3 ~ Solving Two-Step Equations

Solve each equation for the variable. Show all work necessary to justify your answer.

24. $8x - 9 = 7$

25. $\dfrac{h}{4} + 3 = -6$

26. $21 = 7 - 2x$

27. $2m + 6 = 18$

28. $0.5y + 1.9 = 10.4$

29. $-6x + 14 = 2$

Write an equation for each statement. Solve the problem. Show all work necessary to justify your answer.

30. Four times a number x plus eleven is thirty-five. Find the number.

31. The number y divided by negative six decreased by seven is eleven. Find the number.

32. Partridge Middle School hosts an ice cream social each month. This month, there were eight more than three times as many people at the ice cream social than last month. There were 128 people at the ice cream social this month. How many people attended last month's ice cream social? Show all work necessary to justify your answer.

Lesson 4.4 ~ The Distributive Property

Use the Distributive Property to rewrite each expression without parentheses.

33. $8(x + 10)$

34. $3(5y - 4)$

35. $-2(x - 9)$

Solve each equation for the variable. Show all work necessary to justify your answer.

36. $3(x - 5) = 21$

37. $-8(x + 4) = 16$

38. $26 = 2(3x + 1)$

39. $-5(0.3x + 1) = -20$

40. $\frac{1}{2}(8x + 2) = -3$

41. $77 = -11(x - 2)$

42. Three sisters went to the movie theater together on Saturday night. It cost $7.50 for each sister to enter the theater. Each sister spent an equal amount of money at the concession stand. The evening cost the three sisters $37.50.
  a. Let c represent the amount of money each sister spent at the concession stand. Write an equation using the Distributive Property to represent this situation.
  b. How much did each sister spend at the concession stand? Show all work necessary to justify your answer.

Lesson 4.5 ~ Simplifying Expressions

Simplify each algebraic expression.

43. $8 - 7y + 6 + 10y$

44. $11a + 5d - a - 10a - 7d$

45. $5(2x + 7) + x - 4$

46. $2(4x + 1) - 7x + 2$

47. $15 + 3(p + 2)$

48. $13 - 2(x - 6)$

49. It takes Owen 40 minutes to get to his best friend's house. He spends m minutes riding his bike to the bus stop at a rate of 0.2 miles per minute. He gets on the bus and travels the rest of the way at a rate of 0.5 miles per minute. The total distance traveled is represented by the expression $0.2m + 0.5(40 - m)$.
  a. Simplify this expression.
  b. Owen rides his bike for 16 minutes. How far did he ride? Use words and/or numbers to show how you determined your answer.

Lesson 4.6 ~ Simplifying and Solving Equations

Solve each equation for the variable. Show all work necessary to justify your answer.

50. $7a + 2a - 4a = -45$

51. $10y + 9 - 3y + 1 = 59$

52. $-7 = -3(x + 1) + 2x$

53. $2 + \frac{1}{5}m + \frac{3}{10}m + 7 = 5$

54. $-52 = 4(x - 8) - 2(x - 5)$

55. $40 - 2(d + 5) = 20$

Lesson 4.7 ~ Solving Equations With Variables on Both Sides

Solve each equation for the variable. Show all work necessary to justify your answer.

56. $4x + 35 = 11x$ **57.** $5p - 10 = 8p + 14$ **58.** $25 - 9c = c$

59. $0.4h + 1 = 16 - 1.1h$ **60.** $\frac{3}{8}x - 2 = \frac{1}{8}x + 4$ **61.** $11 + 4a = a + 5$

62. An online video game rental site has a Frequent Gamers Plan. It costs $30 to join and each video game rental for members costs $2. Non-members pay a rental fee of $3.50 per rental. Let g represent the total number of video games rented.
   a. Write two expressions, one representing the cost for g rentals by a member and the other representing the cost of g rentals for a non-member.
   b. For what number of rentals would both plans cost the same amount? Use words and/or numbers to show how you determined your answer.

Lesson 4.8 ~ Linear Inequalities

Write an inequality for each graph shown. Use x as the variable.

63. 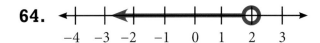 **64.**

65. $5x \geq -20$ **66.** $x + 7 < 6$ **67.** $10 < 3x + 1$

68. $\frac{x}{3} - 2 < -3$ **69.** $19 \geq -5x + 4$ **70.** $9(x + 2) < 6x$

71. Louis has $800 in his bank account at the beginning of the summer. He wants to have at least $500 in his account at the end of the summer. Each week he withdraws $35 for food and entertainment.
   a. Write an inequality for this situation. Let x represent the number of weeks he withdraws money from his account.
   b. How many full weeks can Louis withdraw money from his account?

Tic-Tac-Toe ~ Given an Answer

Write a multi-step inequality that, when solved, is represented by the given inequality.

1. $x > 3$ 2. $x < 5$ 3. $x \geq -4$

4. $x \leq 0$ 5. $x \leq 1.5$ 6. $x > -2.5$

7. $x \geq -8$ 8. $x < 10$ 9. $x \geq \frac{1}{3}$

CAREER FOCUS

RACHELLE
ADOPTION AGENCY DIRECTOR

I am the director for an international adoption agency. I match orphaned children from all over the world with adoptive parents. Some adoption agencies work only with adoptions of children in the Unites States. Other agencies help families adopt children from other countries. I travel to other countries, such as India and Guatemala, to escort children to their new families.

When looking at children who will be coming up for adoption, I look at their growth measurements such as how big they are, how quickly they are growing, and how tall they are. I chart a child's growth on a graph. I use math to determine what percentile they fall in for each measurement. Because I work with countries that use the metric system, I convert metric measurements to customary measurements. This helps parents in the United States better understand how healthy a child is.

A person must have a degree in a related field such as social work or sociology to work on adoptions. It is also a good idea for adoption workers to get some international experience. Visiting other countries helps a person know how that culture works. This is very helpful when adopting from that country.

A person just entering the field of adoptions can expect to make $26,000 - $29,000 a year. That salary varies based on how much experience the person has.

Working as an adoption agent can be very difficult work. Sometimes I come across sad circumstances. However, this job is very rewarding because I can unite a child from a poor situation with a new, loving family.

ACKNOWLEDGEMENTS

**All Photos and Clipart ©2008 Jupiterimages Corporation and Clipart.com
with the exception of the cover photo and the following photos:**

Rational Numbers & Equations Page 3
©iStockphoto.com/Skip ODonnell

Rational Numbers & Equations Page 5
©iStockphoto.com/Ned White

Rational Numbers & Equations Page 6
©iStockphoto.com/Anna Idestam-Almquist

Rational Numbers & Equations Page 14
©iStockphoto.com/david franklin

Rational Numbers & Equations Page 15
©iStockphoto.com/juan moyano

Rational Numbers & Equations Page 16
©iStockphoto.com/Arnel Manalang

Rational Numbers & Equations Page 23
©iStockphoto.com/Robert Simon

Rational Numbers & Equations Page 23
©iStockphoto.com/massimo colombo

Rational Numbers & Equations Page 26
©iStockphoto.com/Apostolos Diamantis

Rational Numbers & Equations Page 29
©iStockphoto.com/Pathathai Chungyam

Rational Numbers & Equations Page 31
©iStockphoto.com/Mark Swallow

Rational Numbers & Equations Page 32
©iStockphoto.com/Eric Isselée

Rational Numbers & Equations Page 33
©iStockphoto.com/Barry Murphy

Rational Numbers & Equations Page 34
©iStockphoto.com/vikif

Rational Numbers & Equations Page 35
©iStockphoto.com/MichaelSvoboda

Rational Numbers & Equations Page 41
©iStockphoto.com/Daniel Bendjy

Rational Numbers & Equations Page 43
©iStockphoto.com/Mati Trommer

Rational Numbers & Equations Page 43
©iStockphoto.com/mediaphotos

Rational Numbers & Equations Page 49
©iStockphoto.com/FotoMak

Rational Numbers & Equations Page 52
©iStockphoto.com/Hedda Gjerpen

Rational Numbers & Equations Page 53
©iStockphoto.com/james steidl

Rational Numbers & Equations Page 61
©iStockphoto.com/Sean Barley

Rational Numbers & Equations Page 62
©iStockphoto.com/Ruth Black

Rational Numbers & Equations Page 65
©iStockphoto.com/CandyBox Images

Rational Numbers & Equations Page 66
©iStockphoto.com/Jani Bryson

Rational Numbers & Equations Page 68
©iStockphoto.com/Jani Bryson

Rational Numbers & Equations Page 69
©iStockphoto.com/Ermin Gutenberger

Rational Numbers & Equations Page 69
©iStockphoto.com/Jon Jakobson

Rational Numbers & Equations Page 75
©iStockphoto.com/Danny Hooks

Rational Numbers & Equations Page 77
©iStockphoto.com/Aldo Murillo

Rational Numbers & Equations Page 77
©iStockphoto.com/John Sigler

Rational Numbers & Equations Page 78
©iStockphoto.com/Roman Milert

Rational Numbers & Equations Page 82
©iStockphoto.com/Goodluz

Rational Numbers & Equations Page 84
©iStockphoto.com/David Freund

Rational Numbers & Equations Page 94
©iStockphoto.com/Wojciech Gajda

Rational Numbers & Equations Page 94
©iStockphoto.com/THEPALMER

Rational Numbers & Equations Page 95
©iStockphoto.com/Scott Smith

Rational Numbers & Equations Page 103
©iStockphoto.com/Nicholas Piccillo

Rational Numbers & Equations Page 115
©iStockphoto.com/Alexander Podshivalov

Rational Numbers & Equations Page 116
©iStockphoto.com/Sergey Dubrovskiy

Rational Numbers & Equations Page 123
©iStockphoto.com/Ljupco

Rational Numbers & Equations Page 130
©iStockphoto.com/g_studio

Rational Numbers & Equations Page 132
©iStockphoto.com/Pgiam

Rational Numbers & Equations Page 137
©iStockphoto.com/Valentyn Volkov

Rational Numbers & Equations Page 141
©iStockphoto.com/Christopher Futcher

Rational Numbers & Equations Page 143
©iStockphoto.com/Maartje van Caspel

Rational Numbers & Equations Page 144
©iStockphoto.com/kristian sekulic

Layout and Design by Judy St. Lawrence

Cover Design by Schuyler St. Lawrence

Glossary Translation by Keyla Santiago and Heather Contreras

CORE FOCUS ON MATH
GLOSSARY ~ GLOSARIO

A

Absolute Value	The distance a number is from 0 on a number line.	Valor Absoluto	La distancia de un número desde el 0 en una recta numérica.
Acute Angle	An angle that measures more than 0° but less than 90°.	Ángulo Agudo	Un ángulo que mide mas 0° pero menos de 90°.
Adjacent Angles	Two angles that share a ray.	Ángulos Adyacentes	Dos ángulos que comparten un rayo.
Algebraic Expression	An expression that contains numbers, operations and variables.	Expresiones Algebraicas	Una expresión que contiene números, operaciones y variables.
Alternate Exterior Angles	Two angles that are on the outside of two lines and are on opposite sides of a transversal.	Ángulos Exteriores Alternos	Dos ángulos que están afuera de dos rectas y están a lados opuestos de una transversal.
Alternate Interior Angles	Two angles that are on the inside of two lines and are on opposites sides of a transversal.	Ángulos Interiores Alternos	Dos ángulos que están adentro de dos rectas y están a lados opuestos de una transversal.
Angle	A figure formed by two rays with a common endpoint.	Ángulo	Una figura formada por dos rayos con un punto final en común.

Area	The number of square units needed to cover a surface.	Área	El número de unidades cuadradas necesitadas para cubrir una superficie.
Ascending Order	Numbers arranged from least to greatest.	Progresión Ascendente	Los números ordenados de menor a mayor.
Associative Property	A property that states that numbers in addition or multiplication expressions can be grouped without affecting the value of the expression.	Propiedad Asociativa	Una propiedad que establece que los números en expresiones de suma o de multiplicación pueden ser agrupados sin afectar el valor de la expresión.
Axes	A horizontal and vertical number line on a coordinate plane. 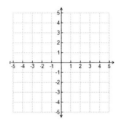	Ejes	Una recta numérica horizontal y vertical en un plano de coordenadas.
Axis of Symmetry	The line of symmetry on a parabola that goes through the vertex. 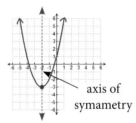 axis of symametry	El Eje De Las Simetría	La linia de simetría de una parábola que pasa por el vértice. 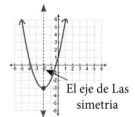 El eje de Las simetria

B

Bar Graph	A graph that uses bars to compare the quantities in a categorical data set. 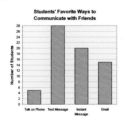	Gráfico de Barras	Una gráfica que utiliza barras para comparar las cantidades en un conjunto de datos categórico. 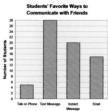
Base (of a power)	The base of the power is the repeated factor. In x^a, x is the base.	Base (de la potencia)	La base de la potenciación es el factor repatidio. En x^a, x es la base.

Glossary ~ Glosario **149**

Base (of a solid)	See Prism, Cylinder, Pyramid and Cone.	Base (de un sólido)	Ver Prisma, Cilindro, Pirámide y Cono.
Base (of a triangle)	Any side of a triangle.	Base (de un triángulo)	Cualquier lado de un triángulo.
Bias	A problem when gathering data that affects the results of the data.	Sesgo	Un problema que ocurre cuando se recogen datos que afectan los resultados de los datos.
Biased Sample	A group from a population that does not accurately represent the entire population.	Muestra Sesgada	Un grupo de una población que no representa con exactitud la población entera.
Binomials	Expressions involving two terms (i.e. $x - 2$).	Binomiales	Expresiones que impliquen dos terminos. (es decir: $x - 2$).
Bivariate Data	Data that describes two variables and looks at the relationship between the two variables.	Datos de dos Variables	Los datos que describen dos variables y analiza la relación entre estas dos variables.
Box-and-Whisker Plot	A diagram used to display the five-number summary of a data set.	Diagrama de Líneas y Bloques	Un diagrama utilizado para mostrar el resumen de cinco números de un conjunto de datos.

C

Categorical Data	Data collected in the form of words.	Datos Categóricos	Datos recopilados en la forma de palabras.
Center of a Circle	The point inside a circle that is the same distance from all points on the circle.	Centro de un Círculo	Un ángulo dentro de un círculo que está a la misma distancia de todos los puntos en el círculo.

Central Angle	An angle in a circle with its vertex at the center of a circle.	Ángulo Central	Un ángulo en un círculo con su vértice en el centro del círculo.
Chord	A line segment with endpoints on a circle.	Cuerda	Un segmento de la recta con puntos finales en el círculo.
Circle	The set of all points that are the same distance from a center point.	Círculo	El conjunto de todos los puntos que están a la misma distancia de un punto central.
Circumference	The distance around a circle.	Circunferencia	La distancia alrededor de un círculo.
Coefficient	The number multiplied by a variable in a term.	Coeficiente	El número multiplicado por una variable en un término.
Commutative Property	A property that states numbers can be added or multiplied in any order.	Propiedad Conmutativa	Una propiedad que establece que los números pueden ser sumados o multiplicados en cualquier orden.
Compatible Numbers	Numbers that are easy to mentally compute; used when estimating products and quotients.	Números Compatibles	Números que son fáciles de calcular mentalmente; utilizado cuando se estiman productos y cocientes.
Complementary Angles	Two angles whose sum is 90°.	Ángulos Complementarios	Dos ángulos cuya suma es de 90°.
Complements	Two probabilities whose sum is 1. Together they make up all the possible outcomes without repeating any outcomes.	Complementos	Dos probabilidades cuya suma es de 1. Juntos crean todos los posibles resultados sin repetir alguno.

Glossary ~ Glosario

English	Definition	Spanish	Definición
Completing the Square	The creation of a perfect square trinomial by adding a constant to an expression in the form $x^2 + bx$.	Terminado el Cuadrado	La creación de un trinomio cuadrado perfecto por adición de una constante a una expresión en la forma $x^2 + bx$.
Complex Fraction	A fraction that contains a fractional expression in its numerator, denominator or both. $$\frac{\frac{3}{4}}{\frac{3}{8}}$$	Fracción Compleja	Una fracción que contiene una expresión fraccionaria en su numerador, el denominador o ambos. $$\frac{\frac{3}{4}}{\frac{3}{8}}$$
Composite Figure	A geometric figure made of two or more geometric shapes.	Figura Compuesta	Una figura geométrica formada por dos o más formas geométricas.
Composite Number	A whole number larger than 1 that has more than two factors.	Número Compuesto	Un número entero mayor que el 1 con más de dos factores.
Composite Solid	A solid made of two or more three-dimensional geometric figures.	Sólido Compuesto	Un sólido formado por dos o más figuras geométricas tridimensionales.
Composition of Transformations	A series of transformations on a point.	Composición de Transformaciones	Una serie de transformaciones en un punto.
Compound Probability	The probability of two or more events occurring.	Compuesto de Probabilidad	La probabilidad de dos o más eventos ocurriendo.
Conditional Frequency	The ratio of the observed frequency to the total number of frequencies in a given category from an experiment or survey.	Frecuencia Condicional	La relación de una frecuencia observada para el número total de frecuencias en una categoría dada del experimento o encuesta.
Cone	A solid formed by one circular base and a vertex.	Cono	Un sólido formado por una base circular y una vértice.
Congruent	Equal in measure.	Congruente	Igual en medida.

Congruent Figures	Two shapes that have the exact same shape and the exact same size. 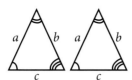	Figuras Congruentes	Dos figuras que tienen exactamente la misma forma y el mismo tamaño. 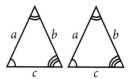
Constant	A term that has no variable.	Constante	Un término que no tiene variable.
Continuous	When a graph can be drawn from beginning to end without any breaks.	Continuo	Cuando una gráfica puede ser dibujada desde principio a fin sin ninguna interrupción.
Conversion	The process of renaming a measurement using different units.	Conversión	El proceso de renombrar una medida utilizando diferentes unidades.
Coordinate Plane	A plane created by two number lines intersecting at a 90° angle. 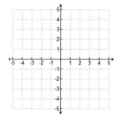	Plano de Coordenadas	Un plano creado por dos rectas numéricas que se intersecan a un ángulo de 90°. 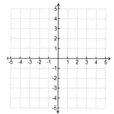
Correlation	The relationship between two variables in a scatter plot.	Correlación	La relación entre dos variables en un gráfico de dispersión.
Corresponding Angles	Two non-adjacent angles that are on the same side of a transversal with one angle inside the two lines and the other on the outside of the two lines. 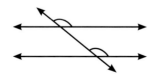	Ángulos Correspondientes	Dos ángulos no adyacentes que están en el mismo lado de una transversal con un ángulo adentro de las dos rectas y el otro afuera de las dos rectas. 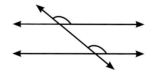
Corresponding Parts	The angles and sides in similar or congruent figures that match.	Partes Correspondientes	Los ángulos y lados en figuras similares o congruentes que concuerdan.

Cube Root	One of the three equal factors of a number. $3 \cdot 3 \cdot 3 = 27 \quad \sqrt[3]{27} = 3$	Raíz Cúbica	Uno de los tres factores iguales de un número. $3 \cdot 3 \cdot 3 = 27 \quad \sqrt[3]{27} = 3$
Cubed	A term raised to the power of 3.	Cubicado	Un término elevado a la potencia de 3.
Cylinder	A solid formed by two congruent and parallel circular bases.	Cilindro	Un sólido formado por dos bases circulares congruentes y paralelas.

D

Decimal	A number with a digit in the tenths place, hundredths place, etc.	Decimal	Un número con un dígito en las décimas, las centenas, etc.
Degrees	A unit used to measure angles.	Grados	Una unidad utilizada para medir ángulos.
Dependent Events	Two (or more) events such that the outcome of one event affects the outcome of the other event(s).	Eventos Dependiente	Dos (o más) eventos de tal manera que el resultado de un evento afecta el resultado del otro evento (s).
Dependent Variable	The variable in a relationship that depends on the value of the independent variable.	Variable Dependiente	La variable en una relación que depende del valor de la variable independiente.
Descending Order	Numbers arranged from greatest to least.	Progresión Descendente	Los números ordenados de mayor a menor.

Diameter	The distance across a circle through the center.	Diámetro	La distancia a través de un círculo por el centro.
Dilation	A transformation which changes the size of the figure but not the shape. 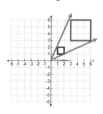	Dilatación	Una transformación que cambia el tamaño de la figura, pero no la forma. 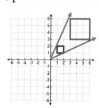
Direct Variation	A linear function that passes through the origin and has the equation $y = mx$.	Variación Directa	Una función lineal que pasa a través del origen y tiene la ecuación $y = mx$.
Discount	The decrease in the price of an item.	Descuento	La disminución de precio en un artículo.
Discrete	When a graph can be represented by a unique set of points rather than a continuous line.	Discreto	Cuando una gráfica puede ser representada por un conjunto de puntos único en vez de una recta continua.
Discriminant	In the quadratic formula, the expression under the radical sign. The discriminant provides information about the number of real roots or solutions of a quadratic equation. $$\frac{-b \pm \sqrt{b^2 - 4ac}}{2a}$$	Discriminante	En la fórmula cuadrática, la expresión bajo el signo radical. El discriminante proporciona información sobre el número o las verdaderas raíces o soluciones de una ecuación cuadrática. $$\frac{-b \pm \sqrt{b^2 - 4ac}}{2a}$$
Distance Formula	A formula used to find the distance between two points on the coordinate plane. $$d = \sqrt{(x_2 - x_1)^2 + (y_2 - y_1)^2}$$	Fórmula de Distancia	Una fórmula utilizada para encontrar la distancia entre dos puntos en un plano de coordenadas. $$d = \sqrt{(x_2 - x_1)^2 + (y_2 - y_1)^2}$$

Glossary ~ Glosario

Distributive Property	A property that can be used to rewrite an expression without parentheses. $a(b + c) = ab + ac$	Propiedad Distributiva	Una propiedad que puede ser utilizada para reescribir una expresión sin paréntesis: $a(b + c) = ab + ac$
Dividend	The number being divided. $100 \div 4 = 25$	Dividendo	El número que es dividido. $100 \div 4 = 25$
Divisor	The number used to divide. $100 \div 4 = 25$	Divisor	El número utilizado para dividir. $100 \div 4 = 25$
Domain	The set of input values of a function.	Dominio	El conjunto de valores entrados de la función.
Dot Plot	A data display that consists of a number line with dots equally spaced above data values. 	Punto de Gráfico	Una visualización de datos que consiste de una línea numérica con puntos igualmente espaciados sobre valores de datos. 
Double Stem-and-Leaf Plot	A stem-and-leaf plot where one set of data is placed on the right side of the stem and another is placed on the left of the stem. 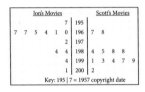	Doble Gráfica de Tallo y Hoja	Una gráfica de tallo y hoja donde un conjunto de datos es colocado al lado derecho del tallo y el otro es colocado a la izquierda del tallo. 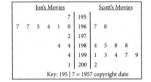

E

Edge	The segment where two faces of a solid meet.	Arista (Borde)	El segmento donde dos caras de un sólido se encuentran.

Elimination Method	A method for solving a system of linear equations.	Método de Eliminación	Un método para resolver un sistema de ecuaciones lineales.
Enlargement	A dilation that creates an image larger than its pre-image.	Agrandamiento	Una dilatación que crea una imagen más grande que su pre-imagen.
Equally Likely	Two or more possible outcomes of a given situation that have the same probability.	Igualmente Probables	Dos o más posibles resultados de una situación dada que tienen la misma probabilidad.
Equation	A mathematical sentence that contains an equals sign between 2 expressions.	Ecuación	Una oración matemática que contiene un símbolo de igualdad entre dos expresiones.
Equiangular	A polygon in which all angles are congruent.	Equiángulo	Un polígono en el cual todos los ángulos son congruentes.
Equilateral	A polygon in which all sides are congruent.	Equilátero	Un polígono en el cual todos los lados son congruentes.
Equivalent Decimals	Two or more decimals that represent the same number.	Decimales Equivalentes	Dos o más decimales que representan el mismo número.
Equivalent Expressions	Two or more expressions that represent the same algebraic expression.	Expresiones Equivalentes	Dos o más expresiones que representan la misma expresión algebraica.
Equivalent Fractions	Two or more fractions that represent the same number.	Fracciones Equivalentes	Dos o más fracciones que representan el mismo número.
Evaluate	To find the value of an expression.	Evaluar	Encontrar el valor de una expresión.
Even Distribution	A set of data values that is evenly spread across the range of the data.	Distribución Igualada	Un conjunto de valores de datos que es esparcido de modo uniforme a través del rango de los datos.

English	Definition	Español	Definición
Event	A desired outcome or group of outcomes.	Evento	Un resultado o grupo de resultados deseados.
Experimental Probability	The ratio of the number of times an event occurs to the total number of trials.	Probabilidad Experimental	La razón de la cantidad de veces que un suceso ocurre a la cantidad total de intentos.
Exponent	In x^a, a is the exponent. The exponent shows the number of times the factor (x) is repeated.	La Exponente	En x^a, a es el exponente. El exponente indica el número de veces que se repite el factor (x).
Exponential Function	A function that can be described by an equation in the form $f(x) = bm^x$.	La Función Exponencial	Una función que puede ser descrito por una ecuasión en la forma $f(x) = bm^x$.

F

English	Definition	Español	Definición
Face	A polygon that is a side or base of a solid.	Cara	Un polígono que es una base de lado de un sólido.
Factors	Whole numbers that can be multiplied together to find a product.	Factores	Números enteros que pueden ser multiplicados entre si para encontrar un producto.
First Quartile (Q1)	The median of the lower half of a data set.	Primer Cuartil (Q1)	Mediana de la parte inferior de un conjunto de datos.
Five-Number Summary	Describes the spread of a data set using the minimum, 1st quartile, median, 3rd quartile and maximum.	Sumario de Cinco Números	Describe la extensión de un conjunto de datos utilizando el mínimo, el primer cuartil, la mediana el tercer cuartil y el máximo.
Formula	An algebraic equation that shows the relationship among specific quantities.	Fórmula	Una ecuación algebraica que enseña la relación entre cantidades específicas.
Fraction	A number that represents a part of a whole number, written as $\frac{\text{numerator}}{\text{denominator}}$.	Fracción	Un número que representa una parte de un número entero, escrito como $\frac{\text{numerador}}{\text{denominador}}$.

Frequency	The number of times an item occurs in a data set.	Frecuencia	La cantidad de veces que un artículo ocurre en un conjunto de datos.
Frequency Table	A table which shows how many times a value occurs in a given interval.	Tabla de Frecuencia	Una tabla que enseña cuantas veces un valor ocurre en un intervalo dado.
Function	A relationship between two variables that has one output value for each input value.	Función	Una relación entre dos variables que tiene un valor de salida para cada valor de entrada.

G

General Form	A quadratic function is in general form when written $f(x) = ax^2 + bx + c$ where $a \neq 0$.	Forma General	Una función cuadrática es en forma general cuándo escrito $f(x) = ax^2 + bx + c$ donde $a \neq 0$.
Geometric Probability	Ratios of lengths or areas used to find the likelihood of an event.	Probabilidad Geométrica	Razones de longitudes o áreas utilizadas para encontrar la probabilidad de un suceso.
Geometric Sequence	A list of numbers that begins with a starting value. Each term in the sequence is generated by multiplying the previous term in the sequence by a constant multiplier.	Secuenciación Geométrica	Una lista de números que comienza con un valor inicial. Cada término de la secuencia se genera al multiplicar el término anterior de la secuencia por un multiplicar constante.
Greatest Common Factor (GCF)	The greatest factor that is common to two or more numbers.	Máximo Común Divisor (MCD)	El máximo divisor que le es común a dos o más números.
Grouping Symbols	Symbols such as parentheses or fraction bars that group parts of an expression.	Símbolos de Agrupación	Símbolos como el paréntesis o barras de fracción que agrupan las partes de una expresión.

H

Height of a Triangle	A perpendicular line drawn from the side whose length is the base to the opposite vertex. 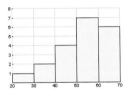	Altura de un Triángulo	Una recta perpendicular dibujada desde el lado cuya longitud es la base al vértice opuesto.
Histogram	A bar graph that displays the frequency of numerical data in equal-sized intervals. 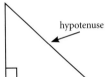	Histograma	Un gráfico de barras que muestra la frecuencia de datos numéricos en intervalos de tamaños iguales. 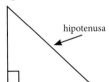
Hypotenuse	The side opposite the right angle in a right triangle.	Hipotenusa	El lado opuesto el ángulo recto en un triángulo rectángulo.

I-J-K

Image	A point or figure which is the result of a transformation or series of transformations.	Imagen	Un punto o figura que es el resultado de una transformación o una serie de transformaciones.
Improper Fraction	A fraction whose numerator is greater than or equal to its denominator.	Fracción Impropia	Una fracción cuyo numerador es mayor o igual a su denominador.
Independent Events	Two (or more) events such that the outcome of one event does not affect the outcome of the other event(s).	Eventos Independientes	Dos (o más) eventos de tal manera que el resultado de un evento no afecta el resultado del otro evento (s).

English	Definition	Spanish	Definición
Independent Variable	The variable representing the input values.	Variable Independiente	La variable que representa los valores entratos.
Inequality	A mathematical sentence using <, >, ≤ or ≥ to compare two quantities.	Desigualdad	Un enunciado matemático usando <, >, ≤ ó ≥ para comparar dos cantidades.
Inference	A logical conclusion based on known information.	Inferencia	Una conclusión lógica basada en la información conocida.
Input-Output Table	A table used to describe a function by listing input values with their output values.	Tabla de Entrada y Salida	Una tabla utilizada para describir una función al enumerar valores de entrada con sus valores de salidas.
Integers	The set of all whole numbers, their opposites, and 0.	Enteros	El conjunto de todos los números enteros, sus opuestos y 0.
Interquartile Range (IQR)	The difference between the 3rd quartile and the 1st quartile in a set of data.	Rango Intercuartil (IQR)	La diferencia entre el tercer cuartil y el primer cuartil en un conjunto de datos.
Inverse Operations	Operations that undo each other.	Operaciones Inversas	Operaciones que se cancelan la una a la otra.
IQR Method	A method for determining outliers using interquartile ranges.	Método IQR	Un método para determinar los datos aberrantes.
Irrational Numbers	A number that cannot be expressed as a fraction of two integers.	Números Irracionales	Un número que no puede ser expresado como una fracción de dos enteros.

Glossary ~ Glosario

Isosceles Trapezoid	A trapezoid that has congruent legs. 	Trapezoide Isósceles	Un trapezoide con catetos congruentes. 
Isosceles Triangle	A triangle that has two or more congruent sides.	Triángulo Isósceles	Un triángulo que tiene dos o más lados congruentes.

L

Lateral Face	A side of a solid that is not a base.	Cara Lateral	Un lado de un sólido que no sea una base.
Least Common Denominator (LCD)	The least common multiple of two or more denominators.	Mínimo Común Denominador (MCD)	El mínimo común múltiplo de dos o más denominadores.
Least Common Multiple (LCM)	The smallest nonzero multiple that is common to two or more numbers.	Mínimo Común Múltiplo (MCM)	El múltiplo más pequeño que no sea cero que le es común a dos o más números.
Leg	The two sides of a right triangle that form a right angle. 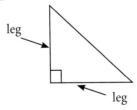	Cateto	Los dos lados de un triángulo rectángulo que forman un ángulo recto. 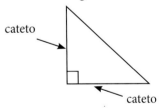
Like Terms	Terms that have the same variable(s).	Términos Semejantes	Términos que tienen el mismo variable(s).

English		Spanish	
Line of Best Fit	A line which best represents the pattern of a two-variable data set.	Recta de Mejor Ajuste	Una recta que mejor representa el patrón de un conjunto de datos de dos variables.
Linear Equation	An equation whose graph is a line.	Ecuación Lineal	Una ecuación cuya gráfica es una recta.
Linear Function	A function whose graph is a line.	Función Lineal	Una función cuya gráfica es una recta.
Linear Pair	Two adjacent angles whose non-common sides are opposite rays.	Par Lineal	Dos ángulos adyacentes cuyos lados no comunes son rayos opuestos.

M

Mark-up	The increase in the price of an item.	Margen de Beneficio	El aumento de precio en un artículo.
Maximum	The highest point on a curve.	Máximo	El punto más alto en la curva.
Mean	The sum of all values in a data set divided by the number of values.	Media	La suma de todos los valores en un conjunto de datos dividido entre la cantidad de valores.
Mean Absolute Deviation	A statistic that shows the average distance from the mean for all numbers in a data set.	Desviación Media Absoluta	Una estadística que muestra la distancia promedio entre la media de todos los números en una serie de datos.

Glossary ~ Glosario **163**

Measures of Center	Numbers that are used to represent a data set with a single value; the mean, median, and mode are the measures of center.	Medidas de Centro	Números que son utilizados para representar un conjunto de datos con un solo valor; la media, la mediana, y la moda son las medidas de centro.
Measures of Variability	Statistics that help determine the spread of numbers in a data set.	Medidas de Variabilidad	Las estadísticas que ayudan a determinar la extensión de los números en una serie de datos.
Median	The middle number or the average of the two middle numbers in an ordered data set.	Mediana	El número medio o el promedio de los dos números medios en un conjunto de datos ordenados.
Minimum	The lowest point on a curve.	Mínimo	El punto más bajo en la curva.
Mixed Number	The sum of a whole number and a fraction less than 1.	Números Mixtos	La suma de un número entero y una fracción menor que 1.
Mode	The number(s) or item(s) that occur most often in a data set.	Moda	El número(s) o artículo(s) que ocurre con más frecuencia en un conjunto de datos.
Motion Rate	A rate that compares distance to time.	Índice de Movimiento	Un índice que compara distancia por tiempo.
Multiple	The product of a number and nonzero whole number.	Múltiplo	El producto de un número y un número entero que no sea cero.

N

Negative Number	A number less than 0.	Número Negativo	Un número menor que 0.

Net	A two-dimensional pattern that folds to form a solid.	Red	Un patrón bidimensional que se dobla para formar un sólido.
Non-Linear Function	A function whose graph does not form a line.	Ecuación No Lineal	Una ecuación cuya gráfica no forma una recta.
Normal Distribution	A set of data values where the majority of the values are located in the middle of the data set and can be displayed by a bell-shaped curve.	Distribución Normal	Un conjunto de valores de datos donde la mayoría de los valores están localizados en el medio del conjunto de datos y pueden ser mostrados por una curva de forma de campana.
Numerical Data	Data collected in the form of numbers.	Datos Numéricos	Datos recopilados en la forma de números.
Numerical Expressions	An expression consisting of numbers and operations that represents a specific value.	Expresiones Numéricas	Una expresión que consta de números y operaciones que representa un valor específico.

O

Obtuse Angle	An angle that measures more than 90° but less than 180°.	Ángulo Obtuso	Un ángulo que mide más de 90° pero menos de 180°.
Opposites	Numbers that are the same distance from 0 on a number line but are on opposite sides of 0.	Opuestos	Números a la misma distancia del 0 en un recta numérica pero en lados opuestos del 0.
Order of Operations	The rules to follow when evaluating an expression with more than one operation.	Orden de Operaciones	Las reglas a seguir cuando se evalúa una expresión con más de una operación.
Ordered Pair	A pair of numbers used to locate a point on a coordinate plane (x, y).	Par Ordenado	Un par de números utilizados para localizar un punto en un plano de coordenadas (x, y).

Origin	The point where the *x*- and *y*-axis intersect on a coordinate plane (0, 0). 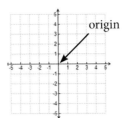	Origen	El punto donde el eje de la *x*-*y* el de la *y*- se cruzan en un plano de coordinadas (0,0). 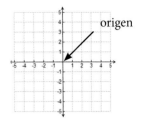
Outcome	One possible result from an experiment or a probability sample space.	Resultado	Un resultado posible de un experimento o un espacio de probabilidad de la muestra.
Outlier	An extreme value that varies greatly from the other values in a data set.	Dato Aberrante	Un valor extremo que varía mucho de los otros valores en un conjunto de datos.

P

Parabola	The graph of a quadratic function. 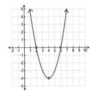	Parábola	La gráfica de una función cuadratica. 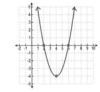
Parallel	Lines in the same plane that never intersect.	Paralela	Rectas en el mismo plano que nunca se intersecan.
Parallel Box-and-Whisker Plot	One box-and-whisker plot placed above another; often used to compare data sets. 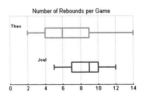	Diagrama Paralelo de Líneas y Bloques	Un diagrama de líneas y bloques ubicado sobre otro para comparar conjuntos de datos. 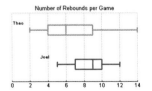

English	Definition	Spanish	Definición
Parallelogram	A quadrilateral with both pairs of opposite sides parallel.	Paralelogramo	Un cuadrilateral con ambos pares de lados opuestos paralelos.
Parent Function	The simplest form of a particular type of function.	Función Principal	La forma más simple de un tipo particular de la función.
Parent Graph	The most basic graph of a function.	Gráfico Matriz	La gráfica más básica de una función.
Percent	A ratio that compares a number to 100.	Por Ciento	Una razón que compara un número con 100.
Percent of Change	The percent a quantity increases or decreases compared to the original amount.	Por Ciento de Cambio	El por ciento que una cantidad aumenta o disminuye comparado a la cantidad original.
Percent of Decrease	The percent of change when the new amount is less than the original amount.	Por Ciento de Disminución	El por ciento de cambio cuando la nueva cantidad es menos que la cantidad original.
Percent of Increase	The percent of change when the new amount is more than the original amount.	Por Ciento de Incremento	El por ciento de cambio cuando la nueva cantidad es más que la cantidad original.
Perfect Cube	A number whose cube root is an integer.	Cubo Perfecto	Un número cuyo raíz cúbica es un número entero.
Perfect Square	A number whose square root is an integer.	Cuadrado Perfecto	Un número cuyo raíz cuadrado es un número entero.
Perfect Square Trinomial	A trinomial that is the square of a binomial.	Trinomio Cuadrado Perfecto	Un trinomio que es el cuadrado de un binomio.
Perimeter	The distance around a figure.	Perímetro	La distancia alrededor de una figura.

Perpendicular	Two lines or segments that form a right angle.	Perpendicular	Dos rectas o segmentos que forman un ángulo recto.
Pi (π)	The ratio of the circumference of a circle to its diameter.	Pi (π)	La razón de la circunferencia de un círculo a su diámetro.
Pictograph	A graph that uses pictures to compare the amounts represented in a categorical data set.	Gráfica Pictórica	Una gráfica que utiliza dibujos para comparar las cantidades representadas en un conjunto de datos categóricos.
Pie Chart	A circle graph that shows information as sectors of a circle.	Gráfico Circular	Enseña la información como sectores de un círculo.
Polygon	A closed figure formed by three or more line segments.	Polígono	Una figura cerrada formada por tres o más segmentos de rectas.
Population	The entire group of people or objects one wants to gather information about.	Población	Todo el grupo de personas o los objetos a los que se quiere obtener información sobre.
Positive Number	A number greater than 0.	Número Positivo	Un número mayor que 0.
Power	An expression such as x^a which consists of two parts, the base (x) and the exponent (a).	Potencia	Una expresión como x^a que consiste de dos partes, la base (x) y el exponente (a).
Pre-image	The original figure prior to a transformation.	Pre-imagen	La figura original antes de una transformación.

Prime Factorization	When any composite number is written as the product of all its prime factors.	Factorización Prima	Cuando cualquier número compuesto es escrito como el producto de todos los factores primos.
Prime Number	A whole number larger than 1 that has only two possible factors, 1 and itself.	Número Primo	Un número entero mayor que 1 que tiene solo dos factores posibles, 1 y el mismo.
Prism	A solid formed by polygons with two congruent, parallel bases.	Prisma	Un sólido formado por polígonos con dos bases congruentes y paralelas.
Probability	The measure of how likely it is an event will occur.	Probabilidad	La medida de cuán probable un suceso puede ocurrir.
Product	The answer to a multiplication problem.	Producto	La respuesta a un problema de multiplicación.
Proper Fraction	A fraction with a numerator that is less than the denominator.	Fracción Propia	Una fracción con un numerador que es menos que el denominador.
Proportion	An equation stating two ratios are equivalent.	Proporción	Una ecuación que establece que dos razones son equivalentes.
Protractor	A tool used to measure angles.	Transportador	Una herramienta para medir ángulos.
Pyramid	A solid with a polygonal base and triangular sides that meet at a vertex.	Pirámide	Un sólido con una base poligonal y lados triangulares que se encuentran en un vértice.

Pythagorean Triple	A set of three positive integers (a, b, c) such that $a^2 + b^2 = c^2$.	Triple de Pitágoras	Un conjunto de tres enteros positivos (a, b, c) tal que $a^2 + b^2 = c^2$.

Q

Q-Points	Points that are created by the intersection of the quartiles for the x- and y-values of a two-variable data set.	Puntos Q	Puntos que son creados por la intersección de los cuartiles para los valores de la x- y la y- de un conjunto de datos de dos variables.
Quadrants	Four regions formed by the x and y axes on a coordinate plane. 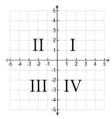	Cuadrantes	Cuatro regiones formadas por el eje-x y el eje-y en un plano de coordenadas. 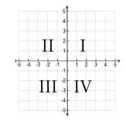
Quadratic Formula	A method which can be used to solve quadratic equations in the form $0 = ax^2 + bx + c$, where $a \neq 0$. $$x = \frac{-b \pm \sqrt{b^2 - 4ac}}{2a}$$	Fórmula Cuadrática	Un método que puede usarse para resolver ecuaciones cuadraticas en la forma $0 = ax^2 + bx + c$, donde $a \neq 0$. $$x = \frac{-b \pm \sqrt{b^2 - 4ac}}{2a}$$
Quadratic Function	Any function in the family with the parent function of $f(x) = x^2$.	Función Cuadrática	Cualquier otra función en la familia con la función principal de $f(x) = x^2$.
Quadrilateral	A polygon with four sides.	Cuadrilateral	Un polígono con cuatro lados.
Quotient	The answer to a division problem.	Cociente	La solución a un problema de división.

R

Radius	The distance from the center of a circle to any point on the circle.	Radio	La distancia desde el centro de un círculo a cualquier punto en el círculo.

Random Sample	A group from a population created when each member of the population is equally likely to be chosen.	Muestra Aleatoria	Un grupo de una población creada cuando cada miembro de la población tiene la misma probabilidad de ser elegido.
Range (of a data set)	The difference between the maximum and minimum values in a data set.	Rango	La diferencia entre los valores máximo y mínimo en un conjunto de datos.
Range (of a function)	The set of output values of a function.	Rango (de una función)	El conjunto de valores salidos de la función.
Rate	A ratio of two numbers that have different units.	Índice	Una proporción de dos números con diferentes unidades.
Rate Conversion	A process of changing at least one unit of measurement in a rate to a different unit of measurement.	Conversión de Índice	Un proceso de cambiar por lo menos una unidad de medición en un índice a una diferente unidad de medición.
Rate of Change	The change in y-values over the change in x-values on a linear graph.	Índice de Cambio	El cambio en los valores de y sobre el cambio en los valores de x en una gráfica lineal.
Ratio	A comparison of two numbers using division. $a:b \quad \frac{a}{b} \quad a \text{ to } b$	Razón	Una comparación de dos números utilizando división. $a:b \quad \frac{a}{b} \quad a \text{ a } b$
Rational Number	A number that can be expressed as a fraction of two integers.	Número Racional	Un número que puede ser expresado como una fracción de dos enteros.
Ray	A part of a line that has one endpoint and extends forever in one direction.	Rayo	Una parte de una recta que tiene un punto final y se extiende eternamente en una dirección.
Real Numbers	The set of numbers that includes all rational and irrational numbers.	Números Racionales	El conjunto de números que incluye todos los números racionales e irracionales.

| Reciprocals | Two numbers whose product is 1. | Recíprocos | Dos números cuyo producto es 1. |

| Rectangle | A quadrilateral with four right angles. | Rectángulo | Un cuadrilátero con cuatro ángulos rectos. |

| Recursive Routine | A routine described by stating the start value and the operation performed to get the following terms. | Rutina Recursiva | Una rutina descrita al exponer el valor del comienzo y la operación realizada para conseguir los términos siguientes. |

| Recursive Sequence | An ordered list of numbers created by a first term and a repeated operation. | Secuencia Recursiva | Una lista de números ordenados creada por un primer término y una operación repetida. |

| Reduction | A dilation that creates an image smaller than its pre-image. | Reducción | Una dilatación que crea una imagen más pequeña que su pre-imagen. |

| Reflection | A transformation in which a mirror image is produced by flipping a figure over a line. | Reflexión | Una transformación en el que se produce una imagen reflejada volteando una figura sobre una línea. |

| Relative Frequency | The ratio of the observed frequency to the total number of frequencies in an experiment or survey. | Frecuencia Relativa | La proporción de la frecuencia observada para el número total de frecuencias en un experimento o estudio. |

| Remainder | A number that is left over when a division problem is completed. | Remanente | Un número que queda cuando un problema de división se ha completado. |

| Repeating Decimal | A decimal that has one or more digits that repeat forever. | Decimal Repetitivo | Un decimal que tiene uno o más dígitos que se repiten eternamente. |

Representative Sample	A group from a population that accurately represents the entire population.	Muestra Representativa	Un grupo de una población que representa con precisión toda la población.
Rhombus	A quadrilateral with four sides equal in measure.	Rombo	Un cuadrilátero con cuatro lados iguales en la medida.
Right Angle	An angle that measures 90°.	Ángulo Recto	Un ángulo que mide 90°.
Roots	The x-intercepts of a quadratic function. 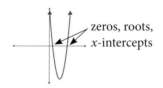 zeros, roots, x-intercepts	Raíces	Las intersecciones-x de una función cuadratica. ceros, raíces, intersecciones-x
Rotation	A transformation which turns a point or figure about a fixed point, often the origin.	Rotación	Una transformación que convierte un punto a una figura sobre un punto fijo. 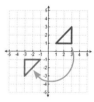

S

Sales Tax	An amount added to the cost of an item. The amount added is a percent of the original amount as determined by a state, county or city.	Impuesto sobre las Ventas	Una cantidad añadida al costo de un artículo. La cantidad añadida es un por ciento de la cantidad original determinado por el estado, condado o ciudad.

Same-Side Interior Angles	Two angles that are on the inside of two lines and are on the same side of a transversal.	Ángulos Interiores del Mismo Lado	Dos ángulos que están en el interior de dos rectas y están en el mismo lado de una transversal.
Sample	A group from a population that is used to make conclusions about the entire population.	Muestra	Un grupo de una población que se utiliza para sacar conclusiones sobre toda la población.
Sample Space	The set of all possible outcomes.	Muestra de Espacio	El conjunto de todos los resultados posibles.
Scale	The ratio of a length on a map or model to the actual object.	Escala	La razón de una longitud en un mapa o modelo al objeto verdadero.
Scale Factor	The ratio of corresponding sides in two similar figures.	Factor de Escala	La razón de los lados correspondientes en dos figuras similares.
Scalene Triangle	A triangle that has no congruent sides.	Triángulo Escaleno	Un triángulo sin lados congruentes.
Scatter Plot	A set of ordered pairs graphed on a coordinate plane.	Diagrama de Dispersión	Un conjunto de pares ordenados graficados en un plano de coordenadas.
Scientific Notation	Scientific notation is an exponential expression using a power of 10 where $1 \leq N < 10$ and P is an integer. $N \times 10^P$	La Notación Científica	Notación científica es una expresión exponencial con una potencia de 10, donde $1 \leq N < 10$ y P es un número entero. $N \times 10^P$

Sector	A portion of a circle enclosed by two radii.	Sector	Una porción de un circulo encerado por dos radios.
Sequence	An ordered list of numbers.	Sucesión	Una lista de números ordenados.
Similar Figures	Two figures that have the exact same shape, but not necessarily the exact same size.	Figuras Similares	Dos figuras que tienen exactamente la misma forma, pero no necesariamente el mismo tamaño exacto.
Similar Solids	Solids that have the same shape and all corresponding dimensions are proportional.	Sólidos Similares	Sólidos con la misma forma y todas sus dimensiones correspondientes son proporcionales.
Simplest Form	A fraction whose numerator and denominator's only common factor is 1.	Expresión Simple	Una fracción cuyo único factor común del numerador y del denominador es 1.
Simplify an Expression	To rewrite an expression without parentheses and combine all like terms.	Simplificar una Expresión	Reescribir una expresión sin paréntesis y combinar todos los términos iguales.
Simulation	An experiment used to model a situation.	Simulación	Un experimento utilizado para modelar una situación.
Single-Variable Data	A data set with only one type of data.	Datos de una Variable	Un conjunto de datos con tan solo un tipo de datos.
Sketch	To make a figure free hand without the use of measurement tools.	Esbozo	Hacer una figura a mano libre sin utilizar herramientas de medidas.
Skewed Left	A plot or graph with a longer tail on the left-hand side.	Torcido a la Izquierda	Un gráfico con una cola al lado izquierdo.

Glossary ~ Glosario **175**

Skewed Right	A plot or graph with a longer tail on the right-hand side.	Torcido a la Derecha	Un gráfico con una cola al lado derecho.
Slant Height	The height of a lateral face of a pyramid or cone.	Altura Sesgada	La altura de un cara lateral de una pirámide o cono.
Slope	The ratio of the vertical change to the horizontal change in a linear graph.	Pendiente	La razón del cambio vertical al cambio horizontal en una gráfica lineal.
Slope Triangle	A right triangle formed where one leg represents the vertical rise and the other leg is the horizontal run in a linear graph.	Triángulo de Pendiente	Un triángulo rectángulo formado donde una cateto representa el ascenso y la otra es una carrera horizontal en una gráfica lineal.
Slope-Intercept Form	A linear equation written in the form $y = mx + b$.	Forma de las Intersecciones con la Pendiente	Una ecuación lineal escrita en la forma $y = mx + b$.
Solid	A three-dimensional figure that encloses a part of space.	Sólido	Una figura tridimensional que encierra una parte del espacio.
Solution	Any value or values that makes an equation true.	Solución	Cualquier valor o valores que hacen una ecuación verdadera.
Solution of a System of Linear Equations	The ordered pair that satisfies both linear equations in the system.	Solución de un Sistema de Ecuaciones Lineales	El par ordenado que satisface ambas ecuaciones lineales en el sistema.

English	Definition	Spanish	Definición
Sphere	A solid formed by a set of points in space that are the same distance from a center point.	Esfera	Un sólido formado por un conjunto de puntos en el espacio que están a la misma distancia de un punto central.
Square	A quadrilateral with four right angles and four congruent sides.	Cuadrado	Un cuadrilátero con cuatro ángulos rectos y cuatro lados congruente.
Square Root	One of the two equal factors of a number. $3 \cdot 3 = 9 \quad 3 = \sqrt{9}$	Raíz Cuadrada	Uno de los dos factores iguales de un número. $3 \cdot 3 = 9 \quad 3 = \sqrt{9}$
Squared	A term raised to the power of 2.	Cuadrado	Un término elevado a la potencia de 2.
Start Value	The output value that is paired with an input value of 0 in an input-output table.	Valor de Comienzo	El valor de salida que es aparejado con un valor de entrada de 0 en una tabla de entradas y salidas.
Statistics	The process of collecting, displaying and analyzing a set of data.	Estadísticas	El proceso de recopilar, exponer y analizar un conjunto de datos.
Stem-and-Leaf Plot	A plot which uses the digits of the data values to show the shape and distribution of the data set.	Gráfica de Tallo y Hoja	Un diagrama que utiliza los dígitos de los valores de datos para mostrar la forma y la distribución del conjunto de datos.
Straight Angle	An angle that measures 180°.	Ángulo Recto	Un ángulo que mide 180°.

Glossary ~ Glosario

Straight Edge	A ruler-like tool with no markings.	Borde Recto	Un gobernante como herramienta sin marcas.
Substitution Method	A method for solving a system of linear equations.	Método de Substitución	Un método para resolver un sistema de ecuaciones lineales.
Supplementary Angles	Two angles whose sum is 180°.	Ángulos Suplementarios	Dos ángulos cuya suma es 180°.
Surface Area	The sum of the areas of all the surfaces on a solid.	Área de la Superficie	La suma de las áreas de todas las superficies en un sólido.
System of Linear Equations	Two or more linear equations.	Sistema de Ecuaciones Lineales	Dos o más ecuaciones lineales.

T

Term	A number or the product of a number and a variable in an algebraic expression; a number in a sequence.	Término	Un número o el producto de un número y una variable en una expresión algebraica; un número en una sucesión.
Terminating Decimal	A decimal that stops.	Decimal Terminado	Un decimal que para.
Theorem	A relationship in mathematics that has been proven.	Teorema	Una relación en las matemáticas que ha sido probada.
Theoretical Probability	The ratio of favorable outcomes to the number of possible outcomes.	Probabilidad Teórica	La proporción de resultados favorables a la cantidad de resultados posibles.
Third Quartile (Q3)	The median of the upper half of a data set.	Tercer Cuartil (Q3)	Mediana de la parte superior de un conjunto de datos.
Tick Marks	Equally divided spaces marked with a small line between every inch or centimeter on a ruler.	Marcas de Graduación	Espacios divididos igualmente marcados con una línea pequeña entre cada pulgada o centímetro en una regla.
Transformation	The movement of a figure on a graph so that it changes size or position.	Transformación	El movimiento de una figura en un gráfico de modo que cambia el tamaño o posición

Translation	A transformation in which a figure is shifted up, down, left or right.	Traducción	Una transformación donde la figura se mudo arriba, abajo, a la izquierda o a la derecha.
Transversal	A line that intersects two or more lines in the same plane.	Transversal	Una recta que interseca dos o más rectas en el mismo plano.
Trapezoid	A quadrilateral with exactly one pair of parallel sides.	Trapezoide	Un cuadrilateral con exactamente un par de lados paralelos.
Tree Diagram	A display that organizes information to determine possible outcomes.	Diagrama de Árbol	Una pantalla que organiza la información para determinar los posibles resulatados.
Trial	A single act of performing an experiment.	Prueba	Un solo intento de realizar un experimento.
Trinomial	An expression with three terms (i.e. $x^2 - 3x + 4$).	Trinomio	Una expreción que tiene tres terminos (es decir: $x^2 - 3x + 4$).
Two-Step Equation	An equation that has two different operations.	Ecuación de Dos Pasos	Una ecuación que tiene dos operaciones diferentes.
Two-Variable Data	A data set where two groups of numbers are looked at simultaneously.	Datos de dos Variables	Un conjunto de datos dónde dos grupos de números se observan simultáneamente.

| Two-Way Frequency Table | A table that shows how many times a value occurs for a pair of categorical data. | Tabla de Frecuencia Bidireccional | Una tabla que muestra cuántas veces aparece un valor de un par de datos categóricos. |

Dog Owners / Walk
	Yes	No
Yes	15	20
No	25	20

Perro Propietario / Paseo
	Si	No
Si	15	20
No	25	20

U-V-W

| Unit Rate | A rate with a denominator of 1. | Índice de Unidad | Un índice con un denominador de 1. |

| Univariate Data | Data that describes one variable (i.e., scores on a test). | Data Univariados | Datos que describen una variable (es decir: puntajes en una prueba). |

| Variable | A symbol that represents one or more numbers. | Variable | Un símbolo que representa uno o más números. |

| Vertex | The minimum or maximum point on a parabola. | Vértice | El mínimo o máximo punto en una parábola. |

| Vertex of a Solid | The point where three or more edges meet. | Vértice de un Sólido | El punto donde tres o más bordes se encuentran. |

| Vertex of a Triangle | A point where two sides of a triangle meet. | Vértice de un Triángulo | Un punto donde dos lados de un triángulo se encuentran. |

Vertex of an Angle	The common endpoint of the two rays that form an angle. 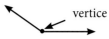	Vértice de un Ángulo	El punto final en común de los dos rayos que forma un ángulo.
Vertex Form	A quadratic function is in vertex form when written $f(x) = a(x - h)^2 + k$ where $a \neq 0$.	Forma De Vértice	Una función cuadrática es en forma general cuándo escrito $f(x) = a(x - h)^2 + k$ donde $a \neq 0$.
Vertical Angles	Non-adjacent angles with a common vertex formed by two intersecting lines. 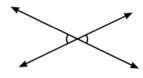	Ángulos Verticales	Ángulos no adyacentes con un vértice en común formado por dos rectas intersecantes.
Vertical Line Test	A test used to determine if a graph represents a function by checking to see if a vertical line passes through no more than one point of the graph of a relation.	Examen Vertical De Línia	Un examen para determinar si una gráfica representa una función. Es utilizada para ver si una línia vertical que pasa a través de no más de un punto de la gráfica de una relación.
Volume	The number of cubic units needed to fill a three-dimensional figure.	Volumen	La cantidad de unidades cúbicas necesitadas para llenar un sólido.

X-Y-Z

x-Axis	The horizontal number line on a coordinate plane. 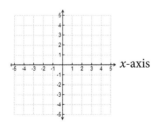	Eje-x, Eje de la x	La recta numérica horizontal en un plano de coordenadas. 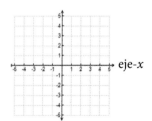

y-Axis	The vertical number line on a coordinate plane. 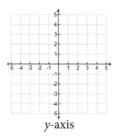 *y*-axis	Eje-*y*, Eje de la *y*	La recta numérica vertical en un plano de coordenadas. 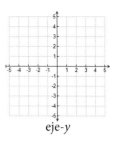 eje-*y*
y-Intercept	The point where a graph intersects the *y*-axis. 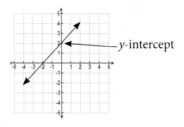	Intersección *y*	El punto donde una gráfica interseca el eje-*y*. 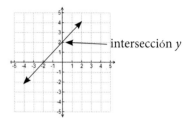
Zero Pair	One positive integer chip paired with one negative integer chip. ● + ● = 0 $1 + (-1) = 0$	Par Cero	Un chip entero positivo emparejado con un chip entero negativo. ● + ● = 0 $1 + (-1) = 0$
Zero Product Property	If a product of two factors is equal to zero, then one or both of the factors must be zero.	Propiedad De Producto Cero	Si un producto de dos factores es iqual a cero, uno o ambos de los factores debe ser cero.
Zeros	The *x*-intercepts of a quadratic function. 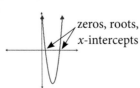	Ceros	Las intersecciones-*x* de una función cuadratica. 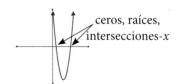

SELECTED ANSWERS

Lesson 1.1

1. Answers may vary. **3.** Answers may vary. **5.** Answers may vary. **7.** 4 **9.** 6 **11.** 20 **13.** $\frac{2}{3}$ **15.** $\frac{1}{3}$ **17.** $\frac{3}{7}$ **19.** $\frac{2}{5}$ **21.** $\frac{2}{3}$ **23.** $\frac{4}{7}$ **25.** $\frac{3}{4}$ and $\frac{3}{4}$ - yes **27.** $\frac{2}{3}$ and $\frac{2}{3}$ - yes **29.** $\frac{3}{8}$ feet **31.** Answers may vary.

Lesson 1.2

1. $\frac{10}{3}$ and $3\frac{1}{3}$ **3.** $\frac{5}{2}$ and $2\frac{1}{2}$ **5.** $\frac{5}{2}$ **7.** $\frac{13}{7}$ **9.** $\frac{19}{4}$ **11.** $\frac{13}{9}$ **13.** $\frac{26}{3}$ **15.** $\frac{65}{12}$ ft **17.** $1\frac{3}{5}$ **19.** $2\frac{3}{4}$ **21.** $4\frac{3}{10}$ **23.** $6\frac{1}{2}$ **25.** $3\frac{3}{4}$ **27.** Mike's hair has grown $\frac{1}{10}$ cm more; See student work for explanation. **29.** $\frac{2}{7}$ **31.** $\frac{1}{3}$ **33.** $\frac{7}{10}$

Lesson 1.3

1. 6 **3.** 36 **5.** 60 **7.** Not correct; no common denominator. $\frac{5}{8}$ **9.** $\frac{3}{5}$ **11.** $\frac{11}{16}$ **13.** $1\frac{1}{6}$ **15.** $\frac{13}{30}$ **17.** $1\frac{5}{12}$ **19.** $1\frac{9}{20}$ **21.** $\frac{3}{16}$ in **23.** $1\frac{5}{12}$ cups **25.** $\frac{11}{15}$; See student work for explanation. **27.** $\frac{11}{5}$ **29.** $5\frac{1}{2}$ **31.** $5\frac{9}{10}$

Lesson 1.4

1. a) ☐ **b)** ■ **c)** ■ **d)** ▨ **e)** $\frac{4}{10}$ **f)** $\frac{2}{5}$ **3.** $\frac{9}{20}$ **5.** $\frac{2}{7}$ **7.** $\frac{1}{9}$ **9.** $\frac{3}{10}$ **11.** $\frac{1}{5}$ **13.** $\frac{2}{5}$ **15.** $\frac{7}{20}$ mile **17.** 20 minutes **19.** 3 **21.** $\frac{3}{8}$ **23.** $2\frac{2}{9}$ **25.** $2\frac{2}{3}$ **27.** $1\frac{1}{7}$ **29.** $\frac{1}{6}$ **31.** 14 ribbons **33.** $\frac{7}{8}$ **35.** $\frac{1}{6}$ **37.** $\frac{1}{8}$ **39.** $1\frac{3}{16}$ pounds

Lesson 1.5

1. Answers may vary. **3.** $5\frac{1}{4}$ **5.** $3\frac{1}{12}$ **7.** $1\frac{1}{2}$ **9.** $9\frac{1}{2}$ **11.** $1\frac{5}{6}$ **13.** $\frac{2}{5}$ **15.** $3\frac{7}{10}$ **17.** $20\frac{1}{8}$ **19.** 8 cups **21.** 6 days; See student work for explanation. **23.** $12\frac{1}{4}$ sq. feet **25.** $1\frac{1}{6}$ **27.** $\frac{1}{22}$ **29.** 0

Lesson 1.6

1. 9.11 **3.** 2.67 **5.** 11.993 **7.** 4.674 **9.** 15.53 **11.** 2.3 **13.** 6.59 pounds **15.** 20.58 miles **17.** $11.98; See student work. **19.** $3\frac{3}{10}$ **21.** $\frac{4}{7}$ **23.** $2\frac{1}{2}$ **25.** $2\frac{7}{8}$

Lesson 1.7

1. 18.6 **3.** 4.88 **5.** 7.54 **7.** 18.09 **9.** 112.32 **11.** $5.41 **13.** 1.9 **15.** 4.3 **17.** 52.1 **19.** $0.60 per pound **21.** $58.22; See student work for explanation. **23.** 0.827 **25.** 18.24 **27.** 133.7

Block 1 Review

1. Answers may vary. **3.** Answers may vary. **5.** 4 **7.** $\frac{1}{2}$ **9.** $\frac{1}{4}$ **11.** $\frac{1}{11}$ **13.** $\frac{5}{6}$ foot **15.** $\frac{5}{3}$ and $1\frac{2}{3}$ **17.** $\frac{13}{3}$ **19.** $\frac{17}{9}$ **21.** $2\frac{1}{4}$ **23.** $12\frac{1}{2}$ **25.** 8 **27.** 30 **29.** $\frac{5}{8}$ **31.** $1\frac{8}{15}$ **33.** $\frac{5}{6}$ **35.** $\frac{7}{24}$ pound; See student work for explanation. **37.** $\frac{1}{14}$ **39.** $\frac{3}{11}$ **41.** $\frac{2}{3}$ **43.** $\frac{3}{5}$ **45.** $1\frac{1}{9}$ **47.** $2\frac{1}{5}$ **49.** 4 boards; See student work for explanation. **51.** 4 **53.** $\frac{5}{7}$ **55.** 3 **57.** $8\frac{7}{10}$ **59.** $5\frac{1}{4}$ cups **61.** 7.88 **63.** 8.17 **65.** 17.11 **67. a)** 13.05 miles **b)** 2.55 miles **69.** 37.8 **71.** 6.2 **73.** 12.5 **75.** $2.38

Lesson 2.1

1. −8 **3.** 7 **5.** 109 **7.** [number line with points at −3, 2, 3] **9.** [number line with points at −9, −6, −3, 0, 3, 6, 9] **11.** Answers may vary. **13.** 1 **15.** $2\frac{1}{4}$ **17.** 6.8 **19.** −12 **21.** −8 **23.** +2 **25.** Answers may vary. **27.** Numbers increase in value from left to right. **29.** > **31.** < **33.** > **35.** −2, −1, 0, 2, 5 **37.** −12, −9, −5, −3, −1 **39.** −21, −17, 0, 17, 21 **41.** Answers may vary. **43.** < **45. a)** No, he did not order the negative numbers correctly. **b)** −10, −7, −3, −1, 0, 1, 4, 9 **47.** $\frac{1}{2}$ **49.** $\frac{13}{24}$ **51.** $\frac{3}{4}$

Lesson 2.2

1. −6 **3.** 1 **5.** −5 **7.** −9 **9.** −1 **11.** −12 **13.** −1 **15.** 22 **17.** 13 **19.** −6 **21.** +$4 **23.** 10 **25.** 23 **27.** −10 **29.** Answers may vary. **31.**

Transaction	Integer
Deposited $60	+60
Withdrew $22	−22
Withdrew $6	−6
Deposited $35	+35
Withdrew $20	−20
Balance	$47

33. −5, −1, 2, 4 **35.** −7, −3, −2, −1, 5 **37.** 12 **39.** $2.50

Lesson 2.3

1. Answers may vary. **3.** 8 **5.** −5 **7.** 10 **9.** 6 **11.** −13 **13.** 1 **15.** 25 **17.** 18 **19.** −13 **21.** 20,320 − (−282) = 20,602 feet **23.** Sometimes true; counterexamples may vary. **25.** Never true; counterexamples may vary. **27.** 9.18 **29.** 3.4 **31.** 17.08

Lesson 2.4

1. Answers may vary. **3.** See student work. **5.** See student work. **7.** −1 **9.** 14 **11.** −55 **13.** 18 **15.** −12 **17.** 12 **19.** 20 **21.** −$35 **23.** −36 feet **25.** always true **27.** always true **29.** 24 **31.** −120 **33.** 18 **35.** −$60 **37.** 10 **39.** −11 **41.** 3 **43.** −5 **45.** 7 + (−5) + 1 = 3

Lesson 2.5

1. Answers may vary. **3.** −4 **5.** 2 **7.** 8 **9.** 8 **11.** 7 **13.** 6 **15.** −4 **17.** −6 **19.** −$15 per week **21.** positive **23.** positive **25.** positive **27.** Rule #2 and 3 are incorrect for multiplication and division. Two negatives multiply to be positive. A positive and a negative number always multiply to be a negative number. **29.** $4\frac{1}{5}$ **31.** 3 **33.** $10\frac{1}{2}$ **35.** $\frac{7}{2}$ **37.** $\frac{23}{4}$ **39.** $\frac{19}{3}$

Lesson 2.6

1. a) −2 **b)** 3 **c)** −8 **3.** 9^4 **5.** 10^2 **7.** 8^5 **9.** positive **11.** negative **13.** positive **15.** (4)(4) = 16 **17.** $\left(\frac{1}{2}\right)\left(\frac{1}{2}\right) = \frac{1}{4}$ **19.** (3)(3)(3)(3) = 81 **21.** (−7)(−7) = 49 **23.** (4)(4)(4)(4)(4) = 1024 **25.** = **27. a)** (5)(5)(5) **b)** 5^3 **c)** 125 **29.** 20 **31.** −8 **33.** 20 **35.** −10 **37.** − 42 **39.** 7

Lesson 2.7

1. 4 **3.** −7 **5.** −90 **7.** −1 **9.** 3 **11.** 0 **13.** −12 **15.** 50
17. $\frac{14+3+4}{3}$, each friend owes $7 **19.** Answers may vary.
21. $(15+1) \cdot 4 - 2 \cdot 3 = 58$ **23.** $(-3)(-3)(-3) = -27$
25. $\left(\frac{1}{2}\right)\left(\frac{1}{2}\right) = \frac{1}{4}$ **27.** $(-1)(-1)(-1)(-1) = 1$ **29.** 63 **31.** −1
33. 11

Block 2 Review

1. −6 **3.** 11 **5.** 7 **7.** 46 **9.** −5 **11.** < **13.** > **15.** −3, −2, 0, 2, 7
17. a)

Week	Activity	Integer
1	Increased by 5 points	+5
2	Dropped 3 points	−3
3	Gained 10 points	+10
4	Fell 7 points	−7
5	Decreased by 6 points	−6

b) −7, −6, −3, 5, 10 **19.** −4 **21.** 21 **23.** 0 **25.** −33 **27.** −$12
29. 8 **31.** −7 **33.** 2 **35.** −8 **37.** 101 **39.** −30 **41.** 81
43. −48 **45.** 45 **47.** −77 **49.** −24° **51.** 7 **53.** −20 **55.** −5
57. −7 **59.** −500 feet per minute **61.** 4^5 **63.** 1^4
65. $(-6)(-6) = 36$ **67.** $(12)(12) = 144$
69. $-(1)(1)(1)(1) = -1$ **71.** = **73.** 27 **75.** 94 **77.** −35
79. a) $4(200) + 10(140)$ **b)** $2,200

Lesson 3.1

1. Answers may vary. **3. a)** false; Answers may vary. **b)** true
5. ≈ 1 **7.** ≈ −6 **9.** ≈ 1 **11.** ≈ 30 **13.** ≈ $16.00 **15.** ≈ −2 kg
17. ≈ 10 **19.** ≈ $-\frac{1}{2}$ **21.** ≈ −3 **23.** ≈ −2 **25.** ≈ 12 pounds
27. Answers may vary; the three values should have an estimated total of 4 pounds. **29.** Answers may vary. **31.** $1\frac{9}{10}$
33. $1\frac{3}{20}$ **35.** $11\frac{3}{4}$ **37.** 10

Lesson 3.2

1. $-\frac{4}{5}$ **3.** $\frac{1}{24}$ **5.** $5\frac{5}{6}$ **7.** $1\frac{7}{10}$ **9.** $-4\frac{1}{6}$ **11. a)** $\frac{3}{4}$ **b)** $-\frac{1}{6}$ foot **13.** 1.5
15. −3.43 **17.** −54.5 **19.** −1.33 **21.** −19.6 **23. a)** 2.4
b) −0.75% **25.** Answers may vary; when a positive and negative number are added, you subtract the smaller value from the larger and take the sign of the larger absolute value.
27. −8, −6, −5, −2 **29.** −9, −8, −7, −4, −1
31. $(-4)(-4)(-4) = -64$ **33.** > **35.** >

Lesson 3.3

1. $\frac{1}{2}$ **3.** $10\frac{3}{4}$ **5.** $-\frac{1}{10}$ **7.** $-5\frac{1}{15}$ **9.** $-\frac{1}{9}$ **11.** $+6\frac{3}{8}$ points **13.** −3.7
15. −1.5 **17.** 0.214 **19.** −10.76 **21.** $128.23 **23.** $\frac{6}{35}$ **25.** $2\frac{1}{3}$

Lesson 3.4

1. a) Answers may vary. **b)** Answers may vary. **3.** ≈ 6 **5.** ≈ 10
7. ≈ 11 **9.** ≈ −5 **11.** ≈ 6 **13.** ≈ 11 days; See student work for explanation. **15.** ≈ 5 cubic feet **17.** Answers may vary.
19. Answers may vary. **21.** 4 **23.** −11 **25.** −10 **27.** 4
29. 46 **31.** 28

Lesson 3.5

1. Answers may vary. **3.** $-\frac{6}{25}$ **5.** $6\frac{1}{4}$ **7.** $-2\frac{3}{4}$ **9.** Answers may vary. **11.** −3.12 **13.** 0.07 **15.** 3.125 **17. a)** 2.6 **b)** 4.5(−2.6)
c) −11.7 cm **19.** 15.2 cm **21.** Answers may vary; numbers can be multiplied in any order. **23.** $-\frac{1}{8}$ **25.** 11 **27.** −1.07
29. To make sure he has enough money, you should round the amounts up. He will not have enough money.

Lesson 3.6

1. a) Answers may vary. **b)** Answers may vary. **3.** $\frac{4}{7}$ **5.** $-1\frac{1}{2}$
7. −13 **9.** 9 members **11.** 176 **13.** 22 **15.** −4.25
17. a) 0.8 and 14.4 **b)** 18 minutes **19. a)** −$21.25 **b)** +$52
21. −2 **23.** 33 **25.** 2 **27.** $\frac{17+2+5}{4}$; $6.00 each

Block 3 Review

1. ≈ 20 **3.** ≈ −1.5 **5.** ≈ 0 **7.** ≈ $19.00 **9.** ≈ 12 **11.** ≈ $-\frac{1}{2}$
13. ≈ −3 **15.** $-\frac{1}{2}$ **17.** −6.51 **19.** $\frac{1}{6}$ **21.** −20.06 **23.** 0.01
25. −$30.83 **27.** $-\frac{3}{4}$ **29.** 15.93 **31.** −2.84 **33.** $-\frac{13}{24}$
35. 10 points **37.** ≈ 3 **39.** ≈ 3 **41.** ≈ 4 **43.** ≈ 6 **45.** ≈ −8
47. ≈ 6 days **49.** −0.35 **51.** $-\frac{1}{11}$ **53.** −0.48 **55.** $-\frac{5}{24}$
57. a) $1\frac{3}{10}$ **b)** $-1\frac{3}{10}\left(5\frac{1}{2}\right)$ **c)** $-7\frac{3}{20}$ feet **59.** $-1\frac{3}{4}$ **61.** 3.3 **63.** $-7\frac{1}{5}$
65. $-4\frac{1}{2}$ **67.** 56 **69.** 14.7 ÷ 0.6; 24.5 minutes

Lesson 4.1

1. $y + 5$ **3.** $7z - 6$ **5.** $2x + 10$ **7.** two less than y **9.** sixty more than twice x **11.** 19 **13.** 25 **15.** $\frac{7}{8}$ **17.** −26
19. a) the number of popcorn containers bought **b)** the number of bottles of juice bought **c)** $53.00 **21.** no **23.** yes
25. no **27.** $x + 6 = 10$; E **29.** $x \div 6 + 4 = 9$; D
31. $\frac{1}{2}x + 9 = 14$; F **33.** −12 **35.** $-2\frac{1}{4}$ **37.** $6\frac{2}{7}$

Lesson 4.2

1. $x = 24$ **3.** $p = 33$ **5.** $k = 8$ **7.** $w = -120$ **9.** $f = 1.2$
11. $x = 12$ **13.** $a = -6$ **15.** $y = 13$ **17.** $4\frac{2}{3}$ years
19. $19 + x = -5$; $x = -24$; See student work for explanation.
21. $x + 12 = 35$; $x = 23$ **23.** $x - 22 = -6$; $x = 16$ **25.** Answers may vary. **27.** 12 **29.** 26 **31.** 18

Lesson 4.3

1. $x = 3$ **3.** $x = 50$ **5.** $y = -3$ **7.** $p = 16$ **9.** $w = 20$
11. $w = -10$ **13.** $x = -9$ **15.** $f = 2$ **17.** $2x + 9 = 25$; $x = 8$
19. $12 - 5w = 52$; $w = -8$ **21. a)** $15 + 12x = 87$; original payment of $15 and $12 for each following payment will total $87 **b)** $x = 6$ months **23.** $22x + 8 = 206$; $x = 9$ birdhouses; See student work for explanation. **25.** $\frac{2}{5}$ **27.** $\frac{2}{15}$
29. $\frac{7}{12}$ **31.** 1 **33.** 7 **35.** 30 **37.** 15 **39.** −12

Lesson 4.4

1. $8x + 40$ **3.** $-6m + 24$ **5.** $-10 - 12p$ **7.** $0.8y + 3.6$
9. $22m - 44$ **11.** $3(3.5) + 3(1.25) = \$14.25$ **13.** $x = -7$
15. $x = 5$ **17.** $x = 0$ **19.** $x = -3\frac{3}{4}$ **21. a)** average daily snowfall
b) $x = 3$ inches **23.** $\frac{1}{4}$ **25.** $\frac{1}{4}$ **27.** $-\frac{7}{8}$

Lesson 4.5

1. a) Answers may vary. **b)** Answers may vary.
3. a) $5y$ and $-8y$ **b)** 1 **c)** $6 + x - 3y$ **5.** $x - y$ **7.** $3m + 41$
9. $2x + 3$ **11.** $4p + 17$ **13.** $6x + 17$ **15.** $11y - 21$ **17.** $3x + 2$
19. a) The number of minutes riding on the bus
b) $30 - 0.35m$ **c)** 26.5 miles **21.** Answers may vary. See student work. **23.** $d = 14.2$ **25.** $x = -9$ **27.** $x = -3$

Lesson 4.6

1. Distribute the 5. Then combine like terms. Subtract 10 from both sides. Then divide both sides by 8. **3.** $y = 5$
5. $w = 4$ **7.** $x = 8$ **9.** $x = 12$ **11.** $w = 12$ **13.** $p = 15$
15. a) $\frac{1}{2}x + 2x = 20$ **b)** $x = 8$ **17. a)** number of minutes spent walking **b)** $m = 50$; James ran for 50 minutes; See student work for explanation. **19.** -53 **21.** 13 **23.** 10
25. $(-4 + 1) \cdot 3 - 2 \cdot 3 = -15$

Lesson 4.7

1. $x = 3$ **3.** $c = 3$ **5.** $y = \frac{1}{2}$ **7.** $h = 3$ **9.** $a = 2.3$ **11.** $w = -3$
13. $x = -2$ **15.** $y = 12$ **17.** $3x + 10 = 5x$; $x = 5$
19. $12 - 2w = 3w - 8$; $w = 4$ **21. a)** Christopher: $4 - 0.3m$; Mom: $0.5m$ **b)** 5 minutes; See student work for explanation.
23. 8 lattes; See student work for explanation. **25.** $y = 5$
27. $y = 56$ **29.** $x = 1$ **31.** $k = 0$ **33.** $1\frac{7}{30}$ **35.** $\frac{11}{12}$ **37.** $-\frac{19}{30}$

Lesson 4.8

1. $x > -2$ **3.** $x \geq 1$
5. $x \geq -4$
7. $x > 3$
9. $x \geq -6$
11. $x \leq 1$
13. $x < -5$
15. $x \leq -2$

17. a) $20x \leq 240$; $x \leq 12$ **b)** $x \leq 5$; See student work to show that 6 boxes would be too many. **19. a)** $4 + 0.35x \leq 10$
b) 17 miles

Block 4 Review

1. $9y$ **3.** $8z - 4$ **5.** 6 **7.** 64 **9.** yes **11.** no **13.** $x = 7$
15. $p = 28$ **17.** $k = 5.5$ **19.** $9x = 36$; $x = 4$
21. $x - 31 = -10$; $x = 21$ **23.** -20; See student work for explanation. **25.** $h = -36$ **27.** $m = 6$ **29.** $x = 2$
31. $\frac{y}{-6} - 7 = 11$; $y = -108$ **33.** $8x + 80$ **35.** $-2x + 18$
37. $x = -6$ **39.** $x = 10$ **41.** $x = -5$ **43.** $14 + 3y$ **45.** $11x + 31$
47. $21 + 3p$ **49. a)** $20 - 0.3m$ **b)** 15.2 miles **51.** $y = 7$
53. $m = -8$ **55.** $d = 5$ **57.** $p = -8$ **59.** $h = 10$ **61.** $a = -2$
63. $x \geq -4$ **65.** $x \geq -4$ **67.** $x > 3$ **69.** $x \geq -3$
71. a) $800 - 35x \geq 500$ **b)** 8 weeks

INDEX

A

Absolute value, 39

Addition
   of decimals, 25
   of fractions, 11
   of integers, 46
   of rational numbers, 80

Algebraic expression
   definition of, 108
   equivalent, 127
   evaluating, 108

Associative Property, 26

B

Base of a power, 60

C

Career Focus
   Adoption Agency Director, 146
   Chef, 36
   Electronics Technician, 105
   Pharmacist, 72

Coefficient, 122

Commutative Property, 26

Comparing integers, 40

Compatible numbers
   definition of, 88
   Explore! In Your Head, 88
   using to estimate, 89

Constant, 122

D

Decimals, 25
   adding, 25
   adding positive and negative, 81
   dividing, 29
   estimating sums and differences, 76
   multiplying, 28
   subtracting, 25
   subtracting positive and negative, 84

Distributive Property, 122

Division
   of decimals, 29
   of fractions, 16
   of integers, 56
   of rational numbers, 97

E

Equation, 109
   one-step, 113
   solution of, 109
   solving with the Distributive Property, 123
   solving by simplifying, 130
      Explore! Equation Manipulation, 130
   solving with variables on both sides, 134
   two-step, 118

Equation mats
   Explore! Introduction to Equation Mats, 113
   Explore! Equation Mats for Two-Step Equations, 118
   one-step equations, 113
   two-step equations, 118

Equivalent expressions, 127

Equivalent fractions, 3
   Explore! Fraction Tiles, 4

Estimation
   Explore! Trip to the Store, 75
   of products and quotients of rational numbers, 88
      Explore! In Your Head, 88
   of sums and differences of rational numbers, 88

Evaluating expressions, 108

Explore!
   Equation Manipulation, 130
   Equation Mats for Two-Step Equations, 118
   Fact Puzzle, 66
   Fraction Tiles, 4
   Fraction Careers, 13
   In Your Head, 88
   Integer Chips, 44
   Introduction to Equation Mats, 113
   Number Jumping, 52
   Positive or Negative?, 60
   Rope Rodeo, 20
   Trip to the Store, 75

Explore!
   What's the difference?, 85
   Where Do I Belong?, 128

Exponent, 60
   squared, 60
   cubed, 60

F

Fractions
   adding, 11
   adding positive and negative, 80
   definition of, 3
   dividing, 16
   equivalent fractions, 3
   estimating sums and differences, 76
   Explore! Fraction Tiles, 4
   Explore! Fraction Careers, 13
   Greatest Common Factor (GCF), 4
   improper fraction, 8
   Least Common Multiple (LCM), 11
   mixed numbers, 8
   multiplying, 15
   reciprocal of, 16
   simplest form, 5
   simplifying, 3
   subtracting, 11
   subtracting positive and negative, 84

G

Greatest Common Factor (GCF), 4

H

I

Improper fraction, 8

Inequalities, 138
   solving and graphing of, 139

Integers, 39
   adding, 44
   comparing, 40
   definition of, 39
   dividing, 56
   multiplying, 52
   negative numbers, 39
   opposites, 39
   ordering, 43
   positive numbers, 39

Integers, 39
   product rules, 92
   subtracting, 49
   zero pair, 44

Inverse operations, 56

J

K

L

Least Common Multiple (LCM), 11

Like terms, 126

Linear inequalities, 138

M

Mixed numbers, 8
   adding, 20
   dividing, 20
   multiplying, 20
   subtraction, 20
   Explore! Rope Rodeo, 20

Multiplication
   of decimals, 28
   of fractions, 15
   of integers, 54
      Explore! Number Jumping, 52
   of rational numbers, 92

N

Negative numbers, 39

O

One-step equation, 113
   Explore! Introduction to Equation Mats, 113

Opposites, 39

Ordering integers, 41

Order of operations, 64
   Explore! Fact Puzzle, 66

P

Positive numbers, 39
Powers, 60
   expanded form, 60
   Explore! Positive or Negative?, 60
   rules for, 61

Properties of Equality, 113

Q

R

Rational numbers, 75
   addition of, 80
   division of, 97
   multiplication of, 92
   subtraction of, 84
      Explore! What's the Difference?, 85

Reciprocal, 16

S

Simplest form, 5

Simplify
   expressions, 126
      Explore! Where Do I Belong?, 128
   fractions, 3
   to solve equations, 130

Solving equations
   using simplifying, 130
      Explore! Equation Manipulation, 130

Solution to an equation, 109

Subtraction
   of decimals, 25
   of fractions, 11
   of integers, 49
   of rational numbers, 84

T

Term, 122
   like terms, 126

Two-step equations, 118
   Explore! Equation Mats for Two-Step Equations, 118

U

V

Variable, 108

W

X

Y

Z

Zero pair, 44
   Explore! Integer Chips, 44

PROBLEM-SOLVING

UNDERSTAND THE SITUATION

- Read then re-read the problem.
- Identify what the problem is asking you to find.
- Locate the key information.

PLAN YOUR APPROACH

Choose a strategy to solve the problem:

- Guess, check and revise
- Use an equation
- Use a formula
- Draw a picture
- Draw a graph
- Make a table
- Make a chart
- Make a list
- Look for patterns
- Compute or simplify

STOP AND THINK

- Did you answer the question that was asked?
- Does your answer make sense?
- Does your answer have the correct units?
- Look back over your work and correct any mistakes.

SOLVE THE PROBLEM

- Use your strategy to solve the problem.
- Show all work.

ANSWER THE QUESTION

- State your answer in a complete sentence.
- Include the appropriate units.

DEFEND YOUR ANSWER

Show that your answer is correct by doing one of the following:

- Use a second strategy to get the same answer.
- Verify that your first calculations are accurate by repeating your process.

SYMBOLS

Algebra and Number Operations

SYMBOL	MEANING
+	Plus or positive
−	Minus or negative
$5 \times n$, $5 \cdot n$, $5n$, $5(n)$	Times (multiplication)
$3 \div 4$, $4\overline{)3}$, $\frac{3}{4}$	Divided by (division)
=	Is equal to
≈	Is approximately
<	Is less than
>	Is greater than
%	Percent
$a : b$ or $\frac{a}{b}$	Ratio of a to b
$5.\overline{2}$	Repeating decimal (5.222…)
≥	Is greater than or equal to
≤	Is less than or equal to
x^n	The nth power of x
(a, b)	Ordered pair where a is the x-coordinate and b is the y-coordinate
±	Plus or minus
$\sqrt{x}$	Square root of x
≠	Not equal to
$x \stackrel{?}{=} y$	Is x equal to y?
$|x|$	Absolute value of x
P(A)	Probability of event A

Geometry and Measurement

SYMBOL	MEANING
≅	Is congruent to
~	Is similar to
∠	Angle
$m\angle$	Measure of angle
△ABC	Triangle ABC
$\overline{AB}$	Line segment AB
$\overrightarrow{AB}$	Ray AB
AB	Length of AB
π	Pi (approximately $\frac{22}{7}$ or 3.14)
°	Degree